AUTOMATIC QUANTUM COMPUTER PROGRAMMING

A Genetic Programming Approach

GENETIC PROGRAMMING SERIES

Series Editor

John Koza
Stanford University

Also in the series:

GENETIC PROGRAMMING AND DATA STRUCTURES: Genetic Programming + Data Structures = Automatic Programming! *William B. Langdon*; ISBN: 0-7923-8135-1

AUTOMATIC RE-ENGINEERING OF SOFTWARE USING GENETIC PROGRAMMING, *Conor Ryan;* ISBN: 0-7923-8653-1

DATA MINING USING GRAMMAR BASED GENETIC PROGRAMMING AND APPLICATIONS, *Man Leung Wong and Kwong Sak Leung;* ISBN: 0-7923-7746-X

GRAMMATICAL EVOLUTION: *Evolutionary Automatic Programming in an Arbitrary Language*, *Michael O'Neill and Conor Ryan;* ISBN: 1-4020-7444-1

GENETIC PROGRAMMING IV: *Routine Human-Computer Machine Intelligence*, *John R. Koza, Martin A. Keane, Matthew J. Streeter, William Mydlowec, Jessen Yu, Guido Lanza;* ISBN: 1-4020-7446-8

GENETIC PROGRAMMING THEORY AND PRACTICE, *edited by Rick Riolo, Bill Worzel;* ISBN: 1-4020-7581-2

AUTOMATIC QUANTUM COMPUTER PROGRAMMING
A Genetic Programming Approach

Lee Spector
Hampshire College

Lee Spector
Hampshire College

Library of Congress Control Number: 2006931640

ISBN-10: 0-387-36496-X
ISBN-13: 978-0387-36496-4
e-ISBN-10: 0-387-36791-8
e-ISBN-13: 978-0387-36791-0

Paperback Edition 2007

9 8 7 6 5 4 3 2 1

springer.com

Contents

Preface

This is a book about the frontiers of computer science that have recently been opened by work in quantum mechanics, but it is also a book about the use of recently developed automatic programming technologies to explore those frontiers. The automatic programming technologies themselves issue from another interdisciplinary frontier of computer science — one born of the intersection of computer science with evolutionary biology. So this is a book about two frontiers of computer science, one being used primarily for the sake of exploring the other.

The selection of topics in this book was made with the intention of showing how genetic programming can be usefully applied to certain problems in quantum computing. To this end, it provides a basic introduction to quantum computing for non-physicists and it also provides a basic introduction to genetic programming for non-computer-scientists. These treatments should be comprehensible to scientifically literate readers who have, at minimum, a passing familiarity with undergraduate-level computer science (e.g. programming concepts) and mathematics (e.g. simple linear algebra). No background in physics is assumed.

Neither the introduction to quantum computing nor the introduction to genetic programming is intended to be comprehensive or even "balanced." Coverage of each field is limited to relatively narrow slices that support the demonstrations found later in the book — those demonstrations show how certain genetic programming techniques can be applied to certain problems in quantum computing. Citations are provided where appropriate to sources that provide more comprehensive and detailed coverage.

The first chapter contains an introduction to quantum computing for non-physicists. The intention is to provide readers with a sense of how quantum computers could possibly deliver the surprising benefits that many researchers envision.

The second chapter details a mathematical (matrix-based) model of quantum computation and describes how this model can be used to simulate quantum computations on classical computers. Such simulation is necessarily inefficient — if we could simulate quantum computers efficiently on classical computers then there'd be little reason to study quantum computing in the first place! But for small computations simulation is indeed possible; this model allows us to use simulation in the "fitness assessment" step of a genetic programming algorithm, described later in the book.

The third chapter describes one particular quantum computer simulation system, the author's QGAME ("Quantum Gate and Measurement Emulator") system, and presents a few of the ways in which quantum programs and quantum computer states can be visually displayed. It concludes with a detailed example of the simulation of a quantum program for Grover's database search problem.

The fourth chapter introduces genetic and evolutionary computation, with a focus on the traditional genetic algorithm. It also discusses, in general terms, the use of parallelism to scale genetic and evolutionary computation technologies up for complex applications, and the applicability of these technologies for various types of problems including those related to quantum computing.

The fifth chapter specializes the treatment of genetic algorithms to genetic *programming*, which is the use of genetic algorithms for automatic programming. It includes a detailed example and a discussion of the steps one must generally take to obtain and understand useful results from a genetic programming system.

The sixth chapter moves beyond traditional genetic programming, and describes the ways in which one can evolve programs that include, for example, multiple data types, modules, and developmental components. Some of these capabilities are particularly useful for the evolution of quantum programs. Emphasis is placed on the author's Push programming language for genetic and evolutionary computation, which provides some of the desired advanced capabilities in unusually simple ways. This chapter concludes with a description of the PushGP genetic programming system, which evolves Push programs, and a brief description of some more radically self-adaptive "autoconstructive evolution" techniques that are enabled by Push.

The seventh and eighth chapters bring the materials from all of the preceding chapters together, first with a discussion of specific strategies for quantum program evolution,[1] and then with concrete examples in which interesting quantum programs were evolved using QGAME, PushGP and related technologies. These examples document a few specific ways in which genetic programming has already helped to explore the power of quantum computing.

The ninth chapter provides a brief summary of the main points of the book and discusses prospects for new discoveries made with the aid of automatic quantum computer programming technologies.

Source code, in Common Lisp, for a minimal version of QGAME is included in the Appendix. Additional related source code is available online from addresses that are cited within the text. Most of these files are also linked to the author's public "code" page.[2]

This book would not have been possible without the close working relationships enjoyed by the author with colleagues and students at Hampshire College in Amherst, Massachusetts. Several of the results that are used as examples in the book emerged from joint work of the author with Herbert J. Bernstein, Howard Barnum, and Nikhil Swamy. Although specific joint results are acknowledged where they occur in the text, these citations do not by themselves fully convey the extent of the influence of these colleagues. Similarly, the novel technologies that are described in the text owe much to the contributions of Chris Perry, Jon Klein, Mark Feinstein, Raymond Coppinger, Alan Robinson, Raphael Crawford-Marks, and Manuel Nickschas. Many of these colleagues also commented on the manuscript of this book, leading to substantial improvements. Additional substantial comments were provided by John Koza, Sameer H. Al-Sakran, and Rennie Nelson. Rebecca S. Neimark provided essential assistance in many phases of the project, including the creation of several of the figures and the design of the cover, which uses an image created by Chris Perry. James Hendler provided critical encouragement and advice, and Leni Bowen and Paula Harmon provided invaluable administrative support.

Some of the materials used in this book derive from those prepared by the author for a series of tutorials on quantum computing presented over several years at the Genetic and Evolutionary Computation Confer-

[1]Note that the term "evolution" is used here and throughout this book in a sense derived from its biological usage: it refers to a process in which a population undergoes variation and natural selection. Some physicists use "evolution" in a more general sense, to describe any change in a system over time. The phrase "quantum program evolution" in this book refers to the generation of quantum programs using techniques derived from biological evolutionary processes.

[2]`http://hampshire.edu/lspector/code.html`

ences (GECCO), for an invited presentation on "Quantum Computation and Artificial Intelligence" at the 1999 National Conference on Artificial Intelligence (AAAI), for a seminar in the Chevron TechNet Advanced Information-Based Modeling seminar series, for a seminar presented at BBN Technologies, and for a course called "Quantum Computing with No Prerequisites of Any Kind" taught at Hampshire College.

This work was supported by a National Science Foundation Director's Award for Distinguished Teaching Scholars, by National Science Foundation grant EIA-0216344, and by the Defense Advanced Research Projects Agency and Air Force Research Laboratory, Air Force Materiel Command, under agreement number F30502-00-2-0611.

Graspings:
wholes and not wholes,
convergent divergent,
consonant dissonant,
from all things one and
from one thing all.
—Heraclitus

Chapter 1

THE POWER OF QUANTUM COMPUTING

This chapter provides a brief, non-technical introduction to quantum computing and outlines both the potential power and the enigmatic nature of quantum computers. It also makes the case for the application of automatic programming technologies to problems in quantum computing, arguing that such technologies can play a unique and important role in the future of this emerging field. The discussion here is general; mathematical and computational details are deferred to later chapters.

1. What is Quantum Computing?

What physical principles govern the processes of computation? Physicists studying this question have recently made a remarkable series of discoveries. These discoveries imply that it may be possible to build *quantum computers* — that is, computers that take advantage of certain quantum mechanical phenomena — that are more powerful, in a fundamental sense, than any other computers previously designed. More than that, they may be more powerful than any other computers previously *imagined*, in the sense that they obey new and more permissive laws of computational complexity.

We use the phrase "quantum computing" to describe computational processes that rely for their efficacy on specifically quantum mechanical properties of information-processing hardware. Of course all computing relies on quantum mechanics in some sense, since quantum mechanics is currently our best theory for describing *all* physical processes. As Rolf Landauer has made clear (Landauer, 1999), "information is inevitably physical," and this means, among other things, that the laws of physics (and in particular the laws of quantum mechanics) underlie all information processing. But as of this writing most information processing can be understood using only *classical* physics and *classical* information theory, and the specifically quantum mechanical properties of the hardware can be ignored. Quantum computing is computing in which the specifically quantum mechanical properties matter a great deal, usually because they are being leveraged to allow us to do things that are not permitted by the classical theories. We call computers that can be understood in terms of the classical theories *classical computers*, and computers that can be understood only in terms of quantum mechanics *quantum computers*.

Why are we interested in quantum computing? One reason is that the size of computing elements continues to shrink at an exponential rate (following "Moore's Law"), with the result that we will be storing bits on devices roughly the size of atoms within the next decade. At these sizes, specifically quantum mechanical effects predominate and we will be doing quantum computing whether we want to or not! But most of the excitement surrounding quantum computing comes not from its inevitability but rather from the discovery that quantum computers can do things beyond the reach of classical computers.

What can quantum computers do that classical computers cannot? This question is still largely open and under active investigation. Indeed, the primary motivation for this book is to provide new tools for the exploration of this question. But we do already know that quantum computers can outperform classical computers in a few specific ways.

At the time of this writing the most spectacular known advantage of quantum over classical computers is the complexity advantage demonstrated by Peter Shor's algorithm for factoring large numbers (Shor, 1994), a problem with practical applications in cryptography and possibly in other areas. Although the classical computational complexity for factoring is not known with certainty, the best known classical factoring algorithms require an amount of time proportional to $2^{n^{\frac{1}{3}} \log(n)^{\frac{2}{3}}}$, where n is the number of digits in the number to be factored. In contrast, Shor's quantum algorithm (Beckman et al., 1996; Shor, 1998) requires time proportional to only $n^2 \log(n) \log\log(n)$. This is an exponential savings, assuming that the known classical factoring algorithms are near optimal (which is not known but is suspected by many to be true).

A rough calculation can give one a feel for just how spectacular this improvement is. Suppose we wish to factor a 5,000 digit number. If we crudely (but consistently) assume that the complexity functions in the previous paragraph are *exact*, that all logarithms are to be taken base 2, and that we can execute one instruction per nanosecond, then we would expect the best known classical algorithms to require about 80 billion years. This is many times the current age of the universe. By contrast, under the same assumptions Shor's quantum factoring algorithm would require less than two seconds! Of course we are ignoring many things that cannot really be ignored in these estimates, such as constants in the complexity functions and the possibility that quantum and classical hardware will support different clock rates. So the exact numbers in these estimates may be quite far off. But the dramatic nature of the exponential speedup is independent of these factors.

A more modest, but also more certain advantage of quantum over classical computers was discovered by Lov Grover. Grover's quantum search algorithm (Grover, 1997) achieves a quadratic speedup over the best classical algorithms for finding a single "marked" item in a database. A classical algorithm must test, on average, half of the n items in the database, while the quantum algorithm can find the item after making only about $\sqrt{n}$ queries. This is less spectacular than the apparent exponential savings of Shor's algorithm, but the quantum complexity advantage is unquestionable, the algorithm has wide applicability, and the savings may be considerable in practice. In Section 3.3 we examine Grover's algorithm in more detail, and in Section 8.2 we demonstrate the use of genetic programming to re-discover an instance of Grover's algorithm.

Can quantum computers speed up other sorts of calculations? Several variants of Shor's and Grover's algorithms have been developed for related problems, and a few other, qualitatively different algorithms have also been discovered.[1] For some types of problems we have also obtained, via mathematical analysis, specific bounds on the possible speedups. But overall our current knowledge is spotty; we know relatively little about what kinds of computations can be sped up, or how much, by the use of quantum hardware. These gaps in our knowledge provide one motivation for the development of technologies that can automatically discover new quantum algorithms.

Quantum computing technology may also provide other kinds of benefits, qualitatively different than those due to the computational complexity advantages that are exemplified by Shor's algorithm and Grover's algorithm. For example, quantum states are "tamper resistant" in a certain sense, and this property can be leveraged to provide secure communication channels upon which it is theoretically *impossible* to eavesdrop. Some of the schemes for such channels require relatively little in the way of quantum hardware engineering, and quantum information technology products for secure communications are already commercially available.

[1]Additional quantum algorithms are described, for example, in (Hogg, 1998; Hogg, 2000; Hallgren, 2002; Hallgren et al., 2003; van Dam and Seroussi, 2002; van Dam et al., 2002).

Other possible technologies may result from the exploitation of phenomena such as quantum superdense coding, quantum teleportation or quantum error correction, allowing information to be moved and/or reconstructed in novel ways. Although these applications and potential applications appear to be quite different in nature from the "speedups" provided by Shor's and Grover's algorithms, *all* of these envisioned applications result from the discovery of quantum algorithms and protocols with novel properties. For this reason they may also benefit from technologies that can automatically discover new quantum algorithms.

Some theorists envision yet further benefits emerging from some future, deeper understanding of quantum mechanics, possibly stemming from new theories of quantum gravity that support new forms of computation. For example, Roger Penrose argues that human consciousness and creativity rely on quantum effects *beyond* those conceivably provided by current models of quantum computation (Penrose, 1989; Penrose, 1997). Regardless of the strength of these arguments — which appear to this author to rest on mistakes about the nature of human cognition — it seems reasonable to expect such exotic forms of quantum computation, if they ever exist, to present challenges to human algorithm designers that are at least as great as those posed by "ordinary" quantum computation. This would further increase the utility of technologies that automatically discover new quantum algorithms.

Several general introductions to quantum computing are available. These include, listed roughly from least to most technical, (Brown, 2000; Milburn, 1997; Brooks, 1999; Rieffel and Polak, 2000; Williams and Clearwater, 1998; Steane, 1998; Gruska, 1999; Nielsen and Chuang, 2000). John Preskill's online lecture notes also provide a comprehensive introduction.[2] Some foundational documents can be found in (Feynman, 1996) and (Hey, 1999). Current research in the quantum computing and quantum information theory is published in a wide range of journals (mostly physics journals) and conference proceedings (such as Shapiro and Hirota, 2003). Many contributions are distributed in pre-print form from the online "arXiv" archive.[3]

2. Possibilities Count

How is it that quantum computers can outperform classical computers? That is, how can the specifically quantum mechanical properties of quantum computing hardware provide non-classical computing power? In Chapter 2 we look at a mathematical characterization of quantum

[2] `http://www.theory.caltech.edu/people/preskill/ph229/`
[3] `http://arxiv.org/archive/quant-ph`

computing that provides, in some sense, the most complete answer to this question. But the mathematical characterization does little to provide *intuitions* about what's *really* going on. Indeed, such intuitions are hard to come by, since better-than-classical quantum algorithms exploit the "weird" aspects of quantum mechanics that have baffled nearly everybody for nearly a century. As Richard Feynman wrote in discussing quantum electrodynamics:

> No, you're not going to be able to understand it. . . . You see, my physics students don't understand it either. That is because I don't understand it. Nobody does. . . . The theory of quantum electrodynamics describes Nature as absurd from the point of view of common sense. And it agrees fully with experiment. So I hope you can accept Nature as She is — absurd. (Feynman, 1985)

The same can be said in regard to some of the deepest questions in quantum computing: we can easily see how the mathematics produces the results, but that's a far cry from understanding *how* or *why* Nature conforms to the particular, counter-intuitive mathematics.

Nevertheless, in this section a brief attempt is made to ground at least *some* fundamental intuitions about how it is that quantum mechanical properties can provide computational advantages.

One perspective on the source of the power of quantum computing is that in quantum computing *possibilities count, even if they never happen.* Furthermore, in well-designed quantum algorithms, each of exponentially many possibilities can be used to perform a part of a computation at the same time.

At first blush this must appear to be a preposterous assertion. How can possibilities that *never happen* influence the outcome of a computation? But there is a sense in which this is literally true, and one can view many if not all of the novel effects of quantum computing as stemming from this fact.

Consider a beam splitter as shown in Figure 1.1, which might be made from a half-silvered mirror. Photons leave the light source on the left, hit the beam splitter in the center, and either reflect to detector A or pass through to detector B. When we turn the light up high, sending out a steady beam, half of the photons are detected at A and half are detected at B. When we turn the light down very low, so that there is only one photon in the system at a time, each photon is detected either at A or at B with 50% probability for each detector. That is, for any given photon there is a possibility that it will be reflected and a possibility that it will pass through. Quantum mechanics tells us that we cannot know which possibility will actually happen — that is, will eventually be detected

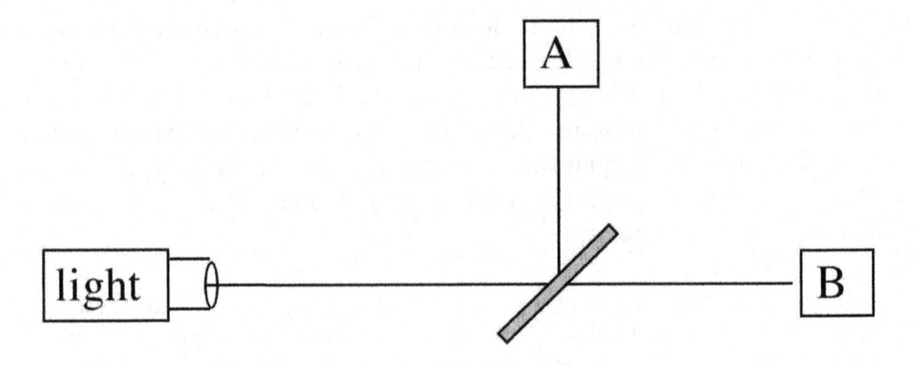

Figure 1.1. A beam splitter.

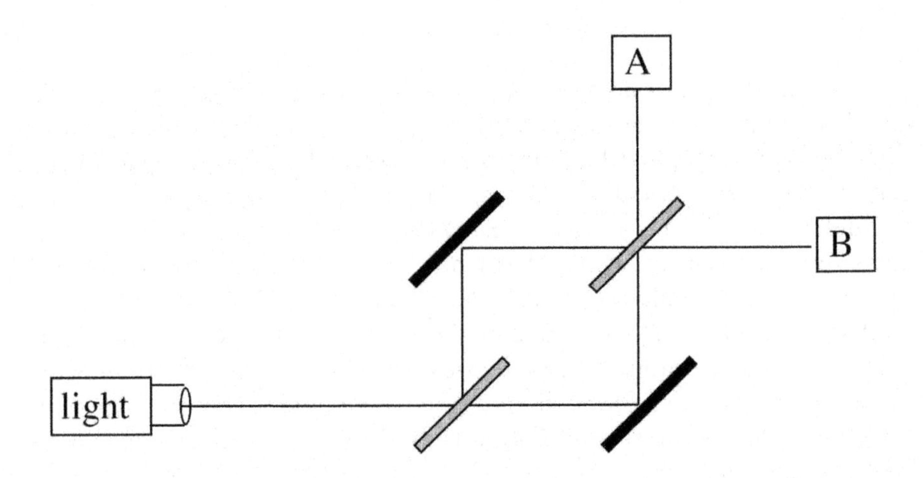

Figure 1.2. An interferometer.

— in advance; both possibilities are "live options" up to the moment of detection.

Now consider the optical interferometer shown in Figure 1.2. The lighter bars are again beam splitters, but the solid black bars are ordinary mirrors. The system is precisely engineered so that each of the four paths in the center of the interferometer is exactly the same length. What happens when we send a beam of photons through this apparatus? One might naively predict that one would again detect half of the photons at A and half at B. After all, our experience with the beam splitter seems to indicate that half of the photons will reflect from the first beam splitter while half will pass through. Each of these beams then reflects back to the second beam splitter where, it would seem reasonable to assume, half of each beam will again reflect and half of each beam will pass

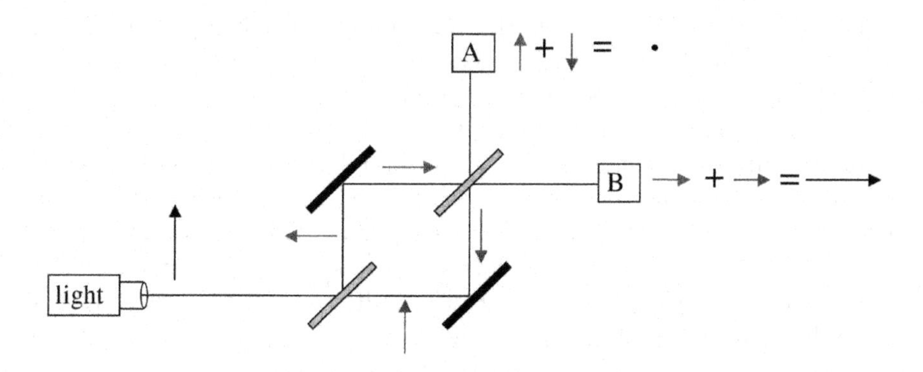

Figure 1.3. Interference of amplitudes in the interferometer.

through. Of the photons that pass through the first splitter, those that pass through the second splitter should be detected at A, while those that reflect from the second splitter should be detected at B. Of the photons that reflect off of the first splitter, those that pass through the second splitter should be detected at B, while those that reflect from the second splitter should be detected at A. The predicted result, overall, would therefore be that we would again detect half of the photons at A and half at B.

But this is not what one actually observes when the experiment is conducted. Instead, *all* of the photons that leave the source are detected at B! How can this be? Perhaps the reader is aware that light has wave-like aspects, along with its particle-like aspects — this may suggest that photons traveling in the different arms of the interferometer interfere with one another, just as waves in a water interfere with one another. Maybe the waves are combining to form higher peaks in some places (such as detector B) and canceling one another in others (such as detector A). This is a reasonable first stab at an explanation, but it begins to break down when we repeat the experiment with the light again turned down very low, so that there is only one photon in the entire apparatus at a time. When we do this we *still* detect photons only at B. Presumably each photon must be taking one path or the other, since nobody has ever detected anything like a "half a photon." What then could be interfering with what?

Quantum mechanics provides a straightforward way to calculate the result that is actually observed, although the interpretation of the calculation — why it is *this* calculation and not some other, and what this *means* about the nature of the universe — is the subject of considerable debate.

The calculation is based on the association of a complex number, called an *amplitude*, with each *possible* path that a photon can take. Graphically, following Feynman (Feynman, 1985), we draw an arrow for each amplitude, as shown in Figure 1.3. The arrow emerging from the light source is of length 1 and is oriented in an arbitrarily chosen direction. Each arrow rotates as its associated photon moves down the path, at a rate related to the photon's frequency. For our interferometer we can ignore these "traveling" rotations by specifying that all of the path segments have lengths that are even multiples of the length required for a full rotation, so that a single arrow can represent the amplitude both at the beginning and at the end of a segment.

When we reach a beam splitter, we split the arrow in two — actually each component is of length $\frac{1}{\sqrt{2}}$, but we can ignore that here — and we rotate the reflected arrow 90° counter-clockwise. We do not rotate the arrow that corresponds to the possibility that the photon passes straight through the beam splitter. At ordinary mirrors the reflected arrows are rotated 180°. When two or more arrows meet via different paths we "add" the arrows using vector addition; that is, we place the arrows tip to tail and draw the "sum" arrow from the tail of the first to the tip of the last. The arrows determine experimental observations in the following way: the square of the length of any arrow gives the *probability* that the photon will be detected by a detector placed in the corresponding path.

This graphical method, based on rotating arrows, can be used to explain a vast array of optical phenomena (as in Feynman, 1985). In the case of our interferometer the explanation emerges quite quickly. Using the rules specified above we see that a full-length arrow emerges at detector B, while no arrow at all emerges at detector A. This means that there is a 100% probability of detecting each photon at detector B, and no probability of detecting a photon at detector A.

In what sense does this demonstrate that "possibilities count"? Consider what happens if we remove the mirror at the lower right of the diagram. In this case there is no possibility of a photon arriving at the second beam splitter via the bottom arm of the interferometer. As a result, the only arrows at the detectors will be those from the upper arm, and a photon arriving via the upper arm will have an equal chance of arriving at either A or B. That is, a single photon leaving the source and traveling on the upper arm may now arrive at A, an outcome that was previously impossible, because the *possibility* of a photon traveling on the lower arm has been eliminated. The possibility of an event that does not occur nonetheless "counts" in determining how a photon in the apparatus will behave.

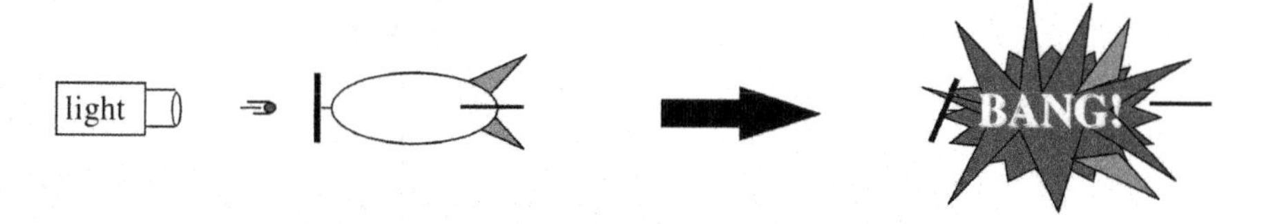

Figure 1.4. A photon-triggered bomb. (Adapted from Penrose, 1997.)

How can this be leveraged for computational advantage? Consider the hypothetical "photon-triggered bomb" illustrated in Figure 1.4. This bomb is fitted with a plunger on its nose, upon which is mounted a mirror. The bomb is designed to detonate if, and only if, a photon hits the mirror. When the bomb detonates, the triggering photon reflects in some direction *other* than that which would result from reflection off of an ordinary mirror. (The specific direction doesn't matter.) Due to a manufacturing error some of the bombs are "duds" of a specific sort — their plungers are stuck, and these dud bombs act as ordinary mirrors. How could we separate the duds from the "good" bombs? The obvious approach of hitting each mirror with a photon has the unfortunate side effect of detonating all of the good bombs. Can we do better?

Avshalom Elitzur and Lev Vaidman (Elitzur and Vaidman, 1993; Vaidman, 1996) discovered how to do this, and their scheme helps to demonstrate how computational work can be done by possibilities that are never actualized. Consider the interferometer in Figure 1.5, in which a photon-triggered bomb has been inserted in place of the mirror at the lower right. First consider what happens when the bomb is a dud. In this case the bomb acts as an ordinary mirror and we have the same situation as in Figure 1.2; all photons leaving the source are detected at B, and none are detected at A. But now consider what happens when the bomb is "good." In this case any photon traveling on the lower arm will detonate the bomb and will fail to reach the second beam splitter. As a consequence, the situation for photons traveling on the upper arm is now the same as it would be with the lower right mirror removed: a single photon leaving the source and traveling on the upper arm may arrive either at A or at B, each with 50% probability. Those that arrive at B tell us nothing — photons would arrive there even if the bomb were a dud. But a photon arriving at A tells us that the bomb must be good. It tells us this by traveling on the upper arm *in a context in which it is not possible to reach the second beam splitter via the lower arm.* We get information (and accomplish computational work) from the presence or absence of possibilities that are not directly explored. The detection of a photon that does not even get close to the bomb tells us that the

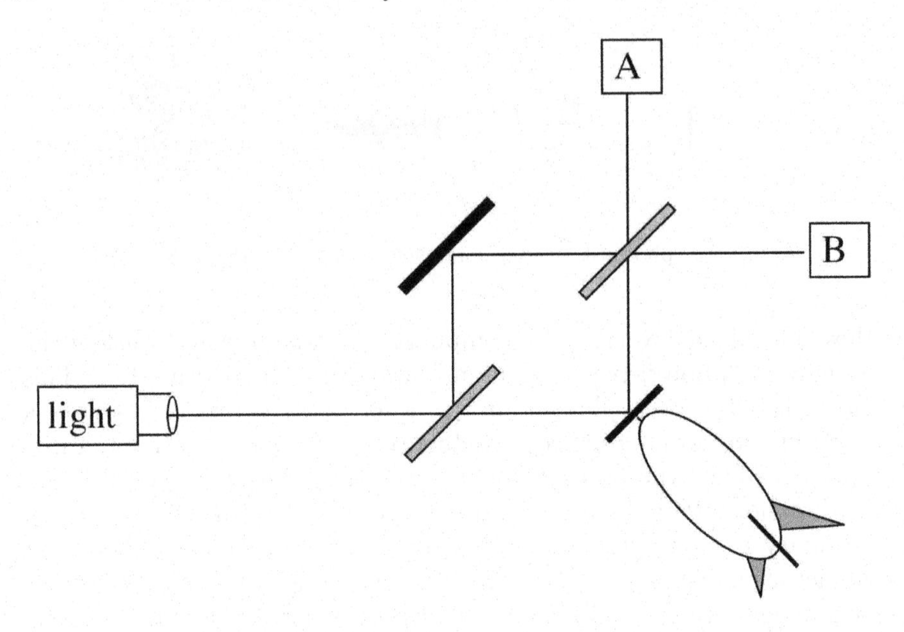

Figure 1.5. A way to test photon-triggered bombs without exploding all of the "good" ones. (Adapted from Penrose, 1997.)

bomb *would* detonate if a photon *were* to strike it. Schemes similar to this can be and have been physically implemented, and while the scheme described here only recovers about a quarter of the good bombs there are enhanced versions that allow one to reduce the amount of bomb loss as much as one would like (Kwiat et al., 1995).[4]

Most interesting quantum algorithms make use of a similar effect. One generally creates a situation in which several possible states of a quantum memory register exist simultaneously, in what is known as "superposition." One then arranges for many of the possibilities to influence, often via some sort of interference, the outcome of later observations. In some cases one can arrange for exponentially many possible computations to simultaneously contribute to the output of a calculation, thereby reducing the amount of time and/or space required to perform a computation below the limits that can be obtained with classical hardware.

3. The Role of Automatic Programming

Computer science will be radically transformed if the ongoing efforts to build large-scale quantum computers eventually succeed and if the

[4]A non-technical discussion of the Elitzur and Vaidman bomb testing problem and its philosophical implications is in (Penrose, 1997, pp. 66–70).

properties of these computers meet optimistic expectations. Unfortunately, however, we still lack a thorough understanding of the power of quantum computing, and it is not always clear how best to utilize the power that we do understand. This is largely because quantum algorithms are difficult to understand and even more difficult to write. Despite large-scale international efforts only a few important quantum algorithms are known, and many basic questions about the potential of quantum algorithms remain unanswered.

Michael Nielsen and Isaac Chuang, in their textbook on *Quantum Computation and Quantum Information*, describe the difficulty of discovering new algorithms as follows:

> Coming up with good quantum algorithms seems to be *hard*. A pessimist might think that's because there's nothing quantum computers are good for other than the applications already discovered! We take a different view. Algorithm design for quantum computers is hard because designers face two difficult problems not faced in the construction of algorithms for classical computers. First, our human intuition is rooted in the classical world. . . . Second, to be truly interesting it is not enough to design an algorithm that is merely quantum mechanical. The algorithm must be *better* than any existing classical algorithm! . . . The combination of these two problems makes the construction of new quantum algorithms a challenging problem for the future. (Nielsen and Chuang, 2000, p. 7)

These circumstances are ideal for the application of automatic programming technologies, which allow us to leverage computer power to explore the space of algorithms in a mechanical way. As mentioned in Section 1 above, such technologies can be applied to the discovery of new quantum speedups and also to the exploitation of other uniquely quantum-computational effects.

Genetic programming techniques, in particular, can be extended to produce quantum algorithms that solve particular computational problems on a quantum computer. These methods have already produced new quantum algorithms and it is reasonable to expect further discoveries in the future. The quantum algorithms found by these methods may help us to understand how to solve particular practical problems using quantum computers. They may also help to guide theoretical work on the power and limits of quantum computing.

The opportunities here are significant both because of the potential power of quantum computing and because of the enigmatic nature of that power. Genetic programming is an automatic programming technology that can, in many circumstances, perform at a "human competitive" level (Koza et al., 2003; see also Section 8.6). But quantum computer programming is particularly difficult for humans, and one might therefore be justified in expecting genetic programming systems to perform *better* than humans in this area. Because the stakes are so high — on account of the unprecedented computational powers that may result from the construction of large scale quantum computers — the application of genetic programming to automatic quantum computer programming is worthy of serious investigation.

Chapter 2

QUANTUM COMPUTER SIMULATION

Chapter 1 discussed quantum computing in non-technical terms and in reference to simple, idealized physical models. In this chapter we make the underlying mathematics explicit and show how one can simulate, albeit inefficiently, the behavior of a quantum computer on an ordinary (classical) digital computer. Such simulation is necessary for the "fitness evaluation" steps of the methods for automatic quantum computer programming that will be described later in this book.

1. Bits, Qubits, and Gates

In classical computing the fundamental unit of information is the *bit*, which can exist in one of two states (conventionally labeled "0" and "1"). Bits can be implemented as positions of gears or switches, levels of charge, or any other conditions of any physical systems that can be easily and unambiguously classified into one of two states. Computations consist of sequences of operations, conventionally referred to as "gates," that are applied to bits and to collections of bits. The physical medium in which the bits and the gates are embedded may influence the computer's size, energy requirements, or "clock rate," but it has no impact on the fundamental computational power of the computer. Two computers with the same storage capacity (in bits) and the same set of supported operations (gates) can be considered equivalent for many purposes.

In quantum computing the fundamental unit of information is the *qubit*, which can also exist in one of two "computational basis" states (conventionally labeled using Paul Dirac's "bra-ket" notation as $|0\rangle$ and $|1\rangle$). But unlike the bit, the qubit can also exist in a *superposition* of $|0\rangle$ and $|1\rangle$ represented as $\alpha_0|0\rangle + \alpha_1|1\rangle$, where α_0 and α_1 are complex numbers such that $|\alpha_0|^2 + |\alpha_1|^2 = 1$. The alphas here are the *ampli-*

tudes described in Chapter 1, and one can square the absolute value of an alpha to determine the probability that a measurement will reveal the corresponding state; for example $|\alpha_0|^2$ is the probability that measurement of the qubit will find it in the $|0\rangle$ state.

The physical medium underlying a quantum computer, like that underlying a classical computer, is generally not relevant to discussions of the computer's fundamental computational power. All that is necessary is that the medium supports the units of storage (qubits) and the relevant operations (quantum gates). A wide variety of proposals has been developed for implementing qubits and quantum gates, including schemes based on optics, ion traps, and the manipulation of nuclear spins in nuclear magnetic resonance devices. All of these schemes present engineering challenges, and many are under active development. We will not be concerned with the details of any of them in the present book, because the computational properties in which we are primarily interested are captured by the abstract view of the quantum computer as a collection of qubits, on which we operate by means of mathematically specified quantum gates. Automatic programming techniques similar to those described later in this book, but built on models of particular implementation schemes, may be be useful for exploring limits or opportunities of the corresponding implementations.

In classical computing the representation of an n-bit system is simply the concatenation of the representations of n 1-bit systems. For example the state of a 5-bit register might be represented as 10010. In quantum computing the representation of a multi-qubit system is more involved, because the individual qubits are not independent of one another. Indeed, qubits in a quantum computer can become "entangled" with one another, and this entanglement underlies several interesting quantum algorithms (Jozsa, 1997; Bennett, 1999). The nature of quantum entanglement is a subject with an enormous literature and a rich history, some of which bears directly on questions about quantum computation. A few suggested entry-points into this literature are (Bell, 1993), (Deutsch, 1997), (Albert, 1992), and many of the essays in (Hey, 1999).

To represent the complete state of a multi-qubit system one must in general store a complex amplitude for each *combination* of basis values ($|0\rangle$ and $|1\rangle$) over the entire system. So, for example, the state of a 3-qubit register might be represented as $\alpha_0|000\rangle + \alpha_1|001\rangle + \alpha_2|010\rangle + \alpha_3|011\rangle + \alpha_4|100\rangle + \alpha_5|101\rangle + \alpha_6|110\rangle + \alpha_7|111\rangle$, where the squares of the absolute values of the alphas sum to 1.

Quantum gates can be formalized as matrices, with the application of a gate to a quantum computer state implemented as the multiplication of the gate's matrix times a column vector containing the state's amplitudes. What sense does this make? Let us first look at a classical version of this idea. Consider a 2-bit classical register. Such a register can be in one of four possible states, namely 00, 01, 10, or 11. Suppose, for reasons that will seem perverse until we generalize to the quantum case, that we wish to represent the state of this register not using the two bits themselves, but rather by recording individually the "amplitudes" for each of the four possible states. Since the register is classical it cannot be in a superposition — it will always be in one particular state. The amplitude corresponding to the actual state of the register will be 1, and all of the other amplitudes will be 0. We will write the amplitudes in the form of a column vector in binary order; that is, the number on top will be the amplitude for the 00 state, the next one will be the amplitude for the 01 state, and so on. So the four possible states of this 2-bit classical register will be represented as:

$$\begin{bmatrix} 1 \\ 0 \\ 0 \\ 0 \end{bmatrix}, \begin{bmatrix} 0 \\ 1 \\ 0 \\ 0 \end{bmatrix}, \begin{bmatrix} 0 \\ 0 \\ 1 \\ 0 \end{bmatrix}, \begin{bmatrix} 0 \\ 0 \\ 0 \\ 1 \end{bmatrix}.$$

What classical operations can be performed on such a register? Although they can be built in various ways from Boolean primitives, all allowable operations have the effect (if they have any effect at all) of changing the state of the register from one of these four states to another. And any such operation can be represented as a matrix, consisting only of 0s and 1s, which, when applied to a state vector (via matrix-vector multiplication), produces another valid state vector. For example, consider the following matrix:

$$\begin{bmatrix} 1 & 0 & 0 & 0 \\ 0 & 1 & 0 & 0 \\ 0 & 0 & 0 & 1 \\ 0 & 0 & 1 & 0 \end{bmatrix}$$

This matrix will have no effect when applied to 00 or 01, but it will transform 10 into 11 and 11 into 10. That is, it will act as a "`NOT`" operation on the right-most bit *if and only if* the left-most bit is 1. For this reason this is often called a "controlled `NOT`" or "`CNOT`" gate. All permissible transformations of the 2-bit register can be represented similarly, using 4×4 matrices containing only 0s and 1s. Not all such matrices are permissible — only those that are guaranteed to produce valid classical state vectors (containing one 1 and the rest 0s) when applied to valid classical state vectors.

Quantum computation can be viewed, mathematically, as a generalization of this classical matrix model. The first generalization is that the amplitudes in the state vectors are no longer required to be 0 or 1. Each amplitude can be any complex number, as long as the squares of the absolute values of the amplitudes sum to 1.[1] Similarly, the set of permissible operations (matrices) is expanded to include any matrix that meets the condition of *unitarity*, which can be expressed (in one formulation) as the requirement that:

$$U^{\dagger}U = UU^{\dagger} = I$$

Here U is the matrix in question, $U^{\dagger}$ is the *Hermitean adjoint* of U (obtained by taking the complex conjugate of each element of U and transposing the result), and I is the identity matrix. The multiplication of a vector of amplitudes by any unitary matrix will always preserve the "summing to one" constraint described above. Although there are infinitely many such unitary matrices, a small finite set suffices for quantum computational *universality* in the same sense that the `NAND` gate suffices for classical computation (Barenco et al., 1995).

In this book we use a selection of quantum gates similar to that used elsewhere in the quantum computing literature. We use the `CNOT` gate described above, along with the simpler 1-bit `Quantum NOT` or `QNOT` gate with the matrix:

$$QNOT \equiv \begin{bmatrix} 0 & 1 \\ 1 & 0 \end{bmatrix}$$

We also use a family of 1-qubit "rotations" parameterized by an angle θ, with matrices of the form:

$$U_{\theta} \equiv \begin{bmatrix} \cos(\theta) & \sin(\theta) \\ -\sin(\theta) & \cos(\theta) \end{bmatrix}$$

Another 1-qubit gate, called `Square Root of NOT` or `SRN` provides a good example of the non-classical power of quantum gates. We use a version of `SRN` with the following matrix (which is also equivalent to $U_{-\frac{\pi}{4}}$):

$$SRN \equiv \frac{1}{\sqrt{2}} \begin{bmatrix} 1 & -1 \\ 1 & 1 \end{bmatrix}$$

[1] Squaring does not obviate the taking of the absolute value, because some amplitudes will be complex and have negative squares.

When applied to a qubit that is in either the $|0\rangle$ or the $|1\rangle$ state, it leaves the qubit in an equal superposition of $|0\rangle$ and $|1\rangle$ — that is, it appears to randomize the value of the qubit, since a measurement after the application of the gate will produce 0 or 1, each with 50% probability. But this is not simple randomization, as the qubit's history can still influence its future behavior. A second application of SRN to the qubit will leave it, deterministically, in the opposite of the state in which it started — that is, measurement will produce 0 if the intial state was $|1\rangle$, or 1 if the initial state was $|0\rangle$.[2] So two applications of SRN produce the effect of QNOT, which is why SRN has the name that it does.

The final 1-qubit gate that we routinely employ is the HADAMARD gate, with the following matrix:

$$H \equiv \frac{1}{\sqrt{2}} \begin{bmatrix} 1 & 1 \\ 1 & -1 \end{bmatrix}$$

This gate is similar to SRN except that it acts more like a "square root of identity." It is useful for creating and "decoding" superpositions in a variety of quantum algorithms.

It is sometimes helpful to use a fully-parameterized 1-qubit gate, which can act as any other 1-qubit gate if its parameters are set appropriately. One form for this "generalized rotation," which we call $U2$, is as follows:

$$U2 \equiv \begin{bmatrix} e^{-i\phi} & 0 \\ 0 & e^{i\phi} \end{bmatrix} \times \begin{bmatrix} \cos(\theta) & \sin(-\theta) \\ \sin(\theta) & \cos(\theta) \end{bmatrix} \times \begin{bmatrix} e^{-i\psi} & 0 \\ 0 & e^{i\psi} \end{bmatrix} \times \begin{bmatrix} e^{i\alpha} & 0 \\ 0 & e^{i\alpha} \end{bmatrix}$$

Other useful 2-qubit gates include the Controlled Phase gates, with matrices of the form:

$$CPHASE \equiv \begin{bmatrix} 1 & 0 & 0 & 0 \\ 0 & 1 & 0 & 0 \\ 0 & 0 & 1 & 0 \\ 0 & 0 & 0 & e^{i\alpha} \end{bmatrix}$$

Finally, the SWAP gate, which simply swaps the states of two qubits, is often handy:

$$SWAP \equiv \begin{bmatrix} 1 & 0 & 0 & 0 \\ 0 & 0 & 1 & 0 \\ 0 & 1 & 0 & 0 \\ 0 & 0 & 0 & 1 \end{bmatrix}$$

[2] Actually, the matrix for two consecutive applications of SRN is $\begin{bmatrix} 0 & -1 \\ 1 & 0 \end{bmatrix}$, meaning that two applications of SRN to $|1\rangle$ will produce $-|0\rangle$, although the change in sign has no effect on measurements. *Six* applications would be required to obtain $|0\rangle$.

Specific problems may call for the use of additional gates. For example, many problems are phrased with respect to a "black box" or "oracle" gate, of which we are asked to determine some property. Grover's database search problem is of this sort; we are given a multi-qubit gate that encodes a database, and we are asked to determine which input will produce a "yes" output (which the oracle usually indicates by flipping — `QNOT`ing — a specified qubit).

2. Gate-Level Simulation

There are many approaches to quantum computer simulation. At one extreme one can attempt to simulate, as realistically as possible, the exact interactions involved in a particular physical device, including noise and other effects of imprecision in the design of the physical components. For example, Kevin Obenland and Alvin Despain simulated a quantum computer that manipulates trapped ions by means of laser pulses, modeling imperfections in the laser apparatus as deviations in the angles of rotations (Obenland and Despain, 1998). Alternatively, one could simulate the quantum computer at a higher level of abstraction, ignoring implementation details and working only with "perfect" unitary matrices.

If one wishes to simulate the execution of *arbitrary* sequences of quantum gates then one necessarily faces exponential space and time costs whether one works at the implementation level or at a more abstract level. That is, if the number of qubits in the system is N, then the space and time requirements for simulation will both scale approximately as 2^N.

In order to evolve quantum algorithms, as described in Chapter 7, we must indeed be able to simulate the execution of arbitrary sequences of quantum gates. But since our focus is on the theoretical power of quantum computing, and not on the strengths or weaknesses of any particular implementation, we can conduct our simulations with straightforward matrix mathematics. We will explicitly maintain full vectors of complex amplitudes, upon which we will explicitly conduct large matrix multiplications. We will pay exponential costs for this form of simulation but the simulation techniques will be conceptually simple.

The exponential costs associated with simulation will limit the range of problems to which our automatic programming techniques can be applied. We will generally seek applications that involve only small quantum systems or that produce algorithms that can be "scaled" to various sizes by hand after they have been discovered automatically. Fortunately, there do seem to be many problems for which the simulation costs are not prohibitive.

Simulation shortcuts are possible if one knows in advance that the algorithm being simulated obeys certain constraints — that is, that certain amplitudes will always be zero, or that certain amplitudes will have values that can be quickly re-derived (so that one needn't always store them all explicitly), or that certain types of entangled states will never be produced. Such constraints, combined with clever encoding schemes, can lead to substantial improvements in simulation speed for many algorithms, although exponential costs will still be incurred in the worst case (Viamontes et al., 2002; Viamontes et al., 2003; Udrescu-Milosav, 2003). These types of advanced simulation techniques are not discussed further in this book, but they could certainly be incorporated into the automatic quantum computer programming framework described here, and one would expect their incorporation to increase the reach of the technology.

To perform the full matrix mathematics described in the previous section we must generally expand the compact matrices that characterize the gates to the appropriate size for the complete quantum system being simulated. For example, if we wish to apply a QNOT gate to the right-most qubit of a 3-qubit system then it is not enough to multiply two amplitudes by the 2×2 matrix that characterizes QNOT. Rather, one must do something that affects *all* amplitudes in the system, effectively multiplying it by the following 8×8 matrix:

$$\begin{bmatrix} 0 & 1 & 0 & 0 & 0 & 0 & 0 & 0 \\ 1 & 0 & 0 & 0 & 0 & 0 & 0 & 0 \\ 0 & 0 & 0 & 1 & 0 & 0 & 0 & 0 \\ 0 & 0 & 1 & 0 & 0 & 0 & 0 & 0 \\ 0 & 0 & 0 & 0 & 0 & 1 & 0 & 0 \\ 0 & 0 & 0 & 0 & 1 & 0 & 0 & 0 \\ 0 & 0 & 0 & 0 & 0 & 0 & 0 & 1 \\ 0 & 0 & 0 & 0 & 0 & 0 & 1 & 0 \end{bmatrix}$$

In this case the expansion of the 2×2 matrix to produce the 8×8 appears relatively straightforward, but the process is more confusing when one must expand a multi-qubit gate, particularly when the qubits to which it is being applied are not adjacent in the chosen representation. For example if one wishes to apply a CNOT gate in a 3-qubit system, using the right-most qubit as the "control" input and the left-most qubit as the "target" (the one that is flipped when the control qubit is 1), then one must effectively use the following 8×8 matrix:

$$\begin{bmatrix} 1 & 0 & 0 & 0 & 0 & 0 & 0 & 0 \\ 0 & 0 & 0 & 0 & 0 & 1 & 0 & 0 \\ 0 & 0 & 1 & 0 & 0 & 0 & 0 & 0 \\ 0 & 0 & 0 & 0 & 0 & 0 & 0 & 1 \\ 0 & 0 & 0 & 0 & 1 & 0 & 0 & 0 \\ 0 & 1 & 0 & 0 & 0 & 0 & 0 & 0 \\ 0 & 0 & 0 & 0 & 0 & 0 & 1 & 0 \\ 0 & 0 & 0 & 1 & 0 & 0 & 0 & 0 \end{bmatrix}$$

How does one construct the needed matrix expansion? First, note that one needn't necessarily construct the matrix (the "tensor product") *explicitly*. In many cases it will suffice to perform an operation which has the same effect as multiplication by the expanded matrix, but which uses only the compact representation of the gate. We call this "implicit matrix expansion."

In other cases one *does* want the explicit representation of the expanded gate, for example because one wants to multiply several expanded gates with one another for storage and later re-application. The choice between implicit and explict matrix expansion presents a trade-off between space requirements and flexibility. With implicit matrix expansion one must store the matrices only in their compact forms, which can be a considerable savings. For example, a 1-qubit gate in its compact form can be represented with only 4 complex numbers, whereas the explicit expansion of this gate for a 10-qubit system consists of $1,048,576$ complex numbers. On the other hand, the expanded forms may be convenient for certain purposes both in the evolution and in the analysis of quantum algorithms. An ideal simulator will therefore provide both options and allow the user to switch among them according to need.

An algorithm for explicit matrix expansion is provided in Figure 2.1, and an algorithm for applying an implicitly expanded gate is provided in Figure 2.2. Source code for these algorithms is included in the distributions of QGAME, a quantum computer programming language and simulation system described in the following chapter; the code for applying an implicitly expanded gate is included in the minimal version of QGAME in the Appendix of this book.

A variety of other approaches to quantum computer simulation exist, some of which are based on alternative conceptualizations of quantum computers (for example, on "quantum Turing machines" or "Feynman computers"). Source code for other simulators can be found in other texts (for example, Williams and Clearwater, 1998) and via internet searches.

To expand gate matrix G (explicitly) for application to an n-qubit system:

- Create a $2^n \times 2^n$ matrix M.
- Let Q be the set of qubit indices to which the operator is being applied, and Q' be the set of the remaining qubit indices.
- $M_{ij} = 0$ if i and j differ from one another, in their binary representations, in any of the positions referenced by indices in Q'.
- Otherwise concatenate bits from the binary representation of i, in the positions referenced by the indices in Q (in numerical order), to produce $i^\star$. Similarly, concatenate bits from the binary representation of j, in the positions referenced by the indices in Q (in numerical order), to produce $j^\star$. Then set $M_{ij} = G_{i^\star j^\star}$.
- Return M.

Figure 2.1. An algorithm for explicit matrix expansion.

To apply gate matrix G (expanded implicitly) to an n-qubit system:

- Let Q be the set of qubit indices to which the operator is being applied, and Q' be the set of the remaining qubit indices.
- For each combination C of 0 and 1 for the set of qubit indices in Q':
  - Extract the column A of amplitudes that results from holding C constant and varying all qubit indices in Q.
  - $A' = G \times A$.
  - Install A' in place of A in the array of amplitudes.

Figure 2.2. An algorithm for applying an implicitly expanded gate.

Chapter 3

QUANTUM COMPUTER PROGRAMMING

This chapter describes the author's QGAME ("Quantum Gate and Measurement Emulator") quantum computer simulation system. It also describes a few of the ways in which quantum programs and quantum computer states can be visually displayed. It concludes with a detailed example of the simulation of a quantum program for an instance of Grover's database search problem.

1. QGAME: Quantum Gate and Measurement Emulator

One embodiment of the simulation ideas presented in Chapter 2 is the author's QGAME system. The original version of QGAME was written in the Common Lisp programming language, which has native support for complex numbers along with other features that support rapid system development, and some of the elements of QGAME's syntax retain Lisp-like features. A C++ version of QGAME, written by Manuel Nickschas, is also available. Current versions of QGAME can be obtained from `http://hampshire.edu/lspector/qgame.html`. Common Lisp source code for the core components of QGAME is provided in the Appendix.

QGAME provides a syntax for the expression of quantum programs and also an interpreter that simulates their execution. Some versions also provide basic visualization capabilities.

A QGAME program consists of a sequence of "instruction expressions," each of which is surrounded by parentheses. The most typical instruction expressions consist of the name of a gate type, followed by a combination of qubit indices (specifying to which qubit or qubits the gate is to be applied) and other parameters (such as angles to rotation gates). For example, an expression of the form:

$$\texttt{(QNOT } q\texttt{)}$$

where q is a qubit index (an integer, starting with 0), applies a quantum "not" (QNOT) gate to the specified qubit. Similarly, an expression of the form:

$$\texttt{(CNOT } q_{control}\ q_{target}\texttt{)}$$

applies a quantum controlled NOT gate to the specified control and target qubits. Instruction expressions following the same pattern, for the remaining gates described in the Chapter 2, are as follows:

$$\texttt{(SRN } q\texttt{)}$$
$$\texttt{(HADAMARD } q\texttt{)}$$
$$\texttt{(U-THETA } q\ \theta\texttt{)}$$
$$\texttt{(U2 } q\ \phi\ \theta\ \psi\ \alpha\texttt{)}$$
$$\texttt{(CPHASE } q_{control}\ q_{target}\ \alpha\texttt{)}$$
$$\texttt{(SWAP } q_{control}\ q_{target}\texttt{)}$$

QGAME also provides a way to specify algorithms that include calls to "oracle" gates with any number of inputs and one output. These gates are "Boolean" in the sense that they can have one of two possible effects on their output qubits on any particular invocation, but unlike classical logic gates they cannot act by *setting* their output qubits to 0 or 1 as such behavior would be non-unitary. The alternative convention adopted in most work on quantum computing, and built into QGAME, is that a Boolean gate acts by *flipping* or *not flipping* its output qubit to indicate an output of 1 or 0 respectively. The "flip" here is implemented as a QNOT, and all oracle gates can therefore be thought of as CNOT gates with more complex controls.

During the testing of an algorithm that contains an oracle gate one normally wants to run the program with various instances of the oracle and to collect statistics over all of the results. For example, if one is testing a program for Grover's search problem one might want to run it on all possible databases (each of which is implemented as an oracle that QNOTs its output qubit if its inputs address the "marked" item), ensuring that it reports the correct answer in each case. This is facilitated in QGAME with a sort of "macro" instruction expression of the form:

$$\texttt{(ORACLE } \Omega\ q_1\ q_2\ \dots\ q_n\ q_{out}\texttt{)}$$

Ω should be the right-hand column of a Boolean truth table that specifies the action of the `ORACLE` gate, listed in parentheses and in binary order. The $q_1, q_2, \ldots q_n$ parameters are the indices of the input qubits, and q_{out} is the index of the output qubit. For example, the following expression:

```
(ORACLE (0 0 0 1) 2 1 0)
```

calls a gate that flips qubit 0 (the right-most qubit) when (and only when) the values of qubits 2 and 1 are both 1. In other words, this oracle acts as the following matrix:

$$\begin{bmatrix} 1 & 0 & 0 & 0 & 0 & 0 & 0 & 0 \\ 0 & 1 & 0 & 0 & 0 & 0 & 0 & 0 \\ 0 & 0 & 1 & 0 & 0 & 0 & 0 & 0 \\ 0 & 0 & 0 & 1 & 0 & 0 & 0 & 0 \\ 0 & 0 & 0 & 0 & 1 & 0 & 0 & 0 \\ 0 & 0 & 0 & 0 & 0 & 1 & 0 & 0 \\ 0 & 0 & 0 & 0 & 0 & 0 & 0 & 1 \\ 0 & 0 & 0 & 0 & 0 & 0 & 1 & 0 \end{bmatrix}$$

This particular matrix, incidentally, is also known as the "Toffoli" gate; it can be used to implement quantum versions of classical `NAND` and `FANOUT` gates, meaning that *all possible* deterministic classical computations can be computed on quantum computers using appropriately connected Toffoli gates (Nielsen and Chuang, 2000, pp. 29–30).

If Ω in an `ORACLE` expression is the symbol `ORACLE-TT` then this indicates that the interpreter should substitute a valid truth table specification in place of the symbol before execution; this is normally in the context of a call to `TEST-QUANTUM-PROGRAM` (see below).

It is sometimes useful to limit the number of times that an oracle can be called during a single simulation. For this reason QGAME also provides an instruction expression of the form:

(`LIMITED-ORACLE` max Ω q_1 q_2 $\ldots$ q_n q_{out})

This works just like `ORACLE` the first *max* times it is executed in a simulation; after *max* executions it has no further effect.

QGAME also provides a way to simulate the effects of single-qubit *measurements* during the execution of a quantum program, and allows for the outcomes of those measurements to influence the remainder of the simulation. In an actual run of a quantum computer such measurements would, in general, be probabilistic. In particular, the probability that measurement of a qubit will find it in the 0 state is equal to the sum of the squares of the absolute values of all of the amplitudes corresponding to 0 values for that qubit. Because we generally wish, when performing our simulations, to obtain the actual probabilities for various outputs and not just particular (probabilistically chosen) outputs, QGAME simulates *all* possible measurement outcomes. This is done by branching the entire simulation and proceeding independently for each possible outcome. In each branch the measured qubit is forced to the measured value. The probability for taking each branch is recorded, and output probabilities at the end of the simulation are calculated on the basis of all possible final states and the probabilities of reaching them.

The syntax for a QGAME measurement is as follows:

`(MEASURE` q`)` $\ldots branch_1 \ldots$ `(END)` $\ldots branch_0 \ldots$ `(END)`

This is actually a sequence of instruction expressions, beginning with the `MEASURE` expression that specifies the qubit to measure. Any number of instruction expressions may occur between the `MEASURE` expression and the first following `END`; all of these will be executed in the branch of the simulation corresponding to a measurement of 1. Similarly, any number of instruction expressions may occur between the first following `END` and a subsequent `END`; all of these will be executed in the branch of the simulation corresponding to a measurement of 0. Instruction expressions following the second `END` will be executed in both branches of the simulation, following the execution of the branch-specific instructions. If there is no `END` following the `MEASURE` expression then the entire remainder of the program is $branch_1$ and there is no $branch_0$. Similarly, if there is only one subsequent `END` then the entire program beyond that `END` is $branch_0$. Unmatched `END`s are ignored.

A few additional instruction expressions provide benefits in special circumstances. Expressions of the form:

`(MATRIX-GATE` M *history*`)`

allow for the inclusion of gates with arbitrary unitary matrices. M here is a fully expanded matrix, of size $2^n \times 2^n$ for an n-qubit system, expressed in Lisp 2D array notation. For example, the notation for a matrix that acts like `QNOT`, for a 1-qubit system, would be "`#2A((0 1)(1 0))`". The *history* parameter is ignored by the QGAME interpreter but it may carry information about the source of the matrix that will be useful for *human* interpretation; this is used, for example, in conjunction with the "gate compression" genetic operator in Chapter 7.

A `HALT` expression simply terminates the current simulation (or the current branch of the simulation, in the context of measurements):

```
(HALT)
```

The following two expressions allow for the printing of diagnostic information:

```
(PRINTAMPS)
(INSP)
```

`PRINTAMPS` prints the amplitudes of the executing quantum system, while `INSP` (short for "inspect") provides more detail about the system state. `INSP` is implementation-specific; in the Lisp version of QGAME it causes the Lisp inspector to be invoked on the executing quantum system, thereby allowing for interactive exploration and manipulation.

The main top-level call to the QGAME interpreter, which will be particularly useful for the approach to automatic quantum computer programming discussed in Chapter 7, is `TEST-QUANTUM-PROGRAM`. This call takes the following inputs:

- `PROGRAM`: The program to be tested, in QGAME program syntax.
- `NUM-QUBITS`: The number of qubits in the quantum computer to be simulated.
- `CASES`: A parenthesized list of "(*oracle-truth-table output*)" pairs, where each *oracle-truth-table* is a parenthesized list of 0s and 1s specifying the right-hand (output) column of the oracle's truth table (where the rows are listed in binary order), and where the *output* is the correct non-negative integer answer for the given truth table; the test compares this number to the number read from the final measurement qubits at the end of the computation.
- `FINAL-MEASUREMENT-QUBITS`: A parenthesized list of indices specifying the qubits upon which final measurements will be performed, with the most significant qubit listed first and the least significant qubit listed last.
- `THRESHOLD`: The probability of error below which a run is considered successful for the sake of the "misses" component of the return value (see below). This is typically set to something like 0.48, which is usually far enough from 0.5 to ensure that the "better than random guessing" performance of the algorithm is not due to accumulated round-off errors.

Additional inputs may be provided by particular implementations to support debugging or other features. `TEST-QUANTUM-PROGRAM` returns a list containing the following values:

- The number of "misses"; that is, cases in which the measured value will, with probability greater than the specified threshold, fail to equal the desired output.
- The maximum probability of error for any provided case.
- The average probability of error for all provided cases.
- The maximum number of expected oracle calls across all cases.
- The number of expected oracle calls averaged across all cases.

It is relatively easy to extend `TEST-QUANTUM-PROGRAM` to return additional values, for example the full list of error values or other statistics related to the program's performance. But the values listed above are sufficient to support many uses of QGAME for automatic quantum computer programming.

2. Visualization

QGAME program syntax provides one way to view quantum algorithms, and lists of amplitudes provide one way to view the state of a quantum computer. But such textual representations, while convenient for computer input and output, are relatively opaque to human comprehension. Alternative visualization techniques can be useful, even in the context of automatic quantum computer programming, as they may significantly aid in the analysis and human understanding of quantum algorithms, whatever their source.

Diagraming schemes similar to those used for classical circuits have been developed for quantum algorithms and they are used frequently in the literature. We use such "gate array" diagrams to document examples later in this book. One typically draws a horizontal line for each qubit and superimposes gate symbols on the lines, indicating from left to right the sequence of gate applications as the computation proceeds across the page. A labeled box is superimposed on a line to represent the application of a single-qubit gate, and boxes or other symbols that span multiple lines are used to represent multi-qubit gates. Our particular diagraming conventions will be made clear in the context of examples.

Gate array diagrams can be helpful, but they can also be deceptive, particularly if one is accustomed to classical circuit or flow diagrams. For example, one must bear in mind that qubits can be entangled, and that gates are really applied not to independent "wires," as implied by the horizontal lines, but rather to amplitudes that are shared among all qubits. Even single-qubit gates typically change *all* of the amplitudes in the system, and the value of every qubit is influenced by every amplitude.

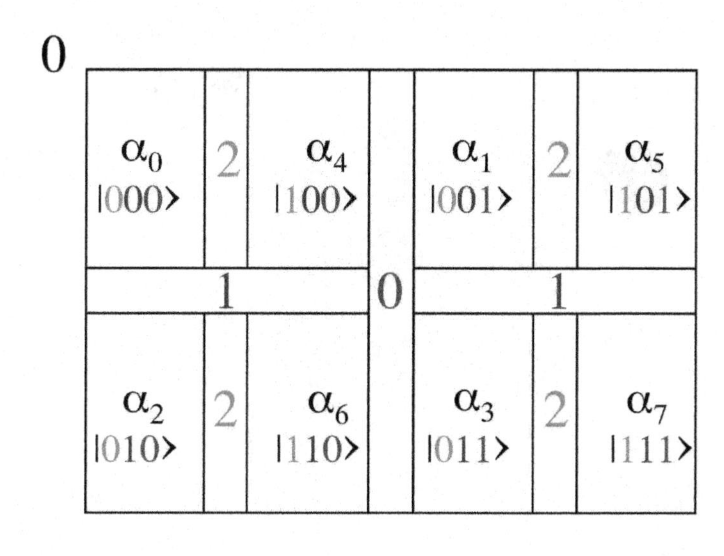

Figure 3.1. An amplitude diagram for a 3-qubit state.

So there may often be interconnections that are essential to the algorithm but are not indicated directly in the diagram.

Quantum algorithms are also often presented in algebraic form. Algebraic representations are also popular for the representation of quantum computer *states*, typically using Dirac "bra-ket" notation. These notations often allow for elegant presentation of algorithms and states designed by *humans*, but they can be ungainly when applied to the arbitrary algorithms and states that emerge from an automatic quantum computer programming system.

Visualization of arbitrary quantum computer states is difficult for several reasons. The state of an n-qubit system is a collection of 2^n amplitudes, each of which is a complex number. The collection of amplitudes has structure, but the structure is n-dimensional and it is not obvious how to map these dimensions onto a 2-dimensional diagram in a meaningful way. Neither is it obvious how best to map the individual complex numbers to image features.

An *amplitude diagram*, as shown in Figure 3.1, can sometimes be helpful. The diagram displays all of the amplitudes numerically (in place of the αs in the Figure), but they are arranged in a grid that hierarchically prioritizes the qubits. In the diagram as shown in Figure 3.1 qubit 0 is prioritized first, so that the major left/right split of the

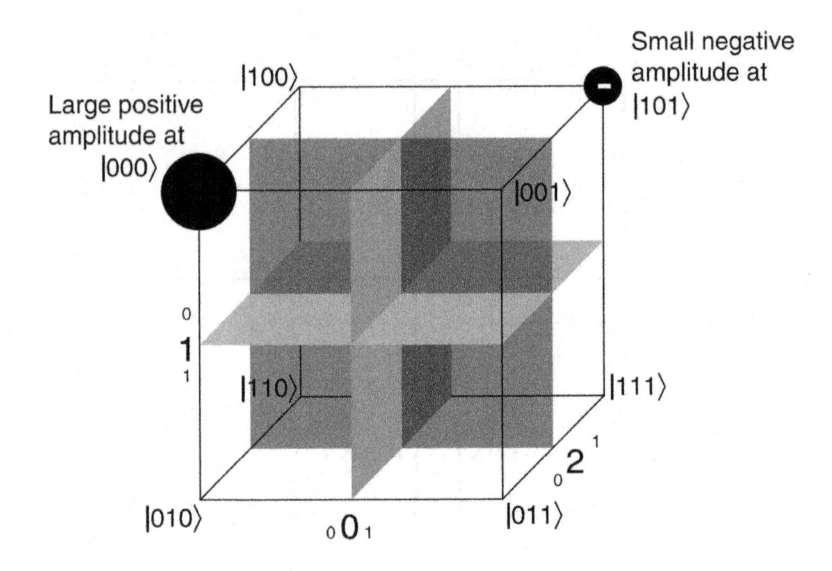

Figure 3.2. A cube diagram for a 3-qubit state.

diagram puts all of the amplitudes for qubit 0 being 0 on the left, and all the amplitudes for qubit 0 being 1 on the right. The second (vertical) split is on the basis of qubit 1, and the third (horizontal, within the quadrants) is on the basis of qubit 2. This particular prioritization makes the values of qubits 0 and 1 most obvious, but one can re-prioritize to focus on other qubits. This scheme can be nested further, allowing for the "hierarchically flattened" visualization of relatively large quantum computer states.

Janet Wiles and Bradley Tonkes developed a similar graphical representation scheme but for rather different (non-quantum) purposes, called *hyperspace graph paper* (Wiles and Tonkes, 2002). Their scheme omits explicit indications of the prioritization of the dimensions (though these could be added), and it uses grayscale values rather than numerals in the individual cells. As a result, many significant patterns are visually evident. For use in visualizing quantum computer states one would have to generalize the grayscale tones in some way, as the amplitudes can be complex. One way to do this, which was adopted in an early graphical user interface for QGAME, is to use a "hue, saturation, value" representation for color, mapping the phase of the amplitude to hue and the absolute value to saturation and/or value.

For the special case of a 3-qubit system a related cube diagram can be used, as shown in Figure 3.2.

3. Example: Grover's Database Search Algorithm

In this section we present one example in detail, an instance of Grover's database search algorithm (Grover, 1997), in order to clarify some of the ideas presented in this chapter.

In this problem we are given a 3-qubit gate that, we are told, implements a 4-item database. Two of the inputs to the gate are used to set an address, and the gate operates by flipping (QNOTing) the third input if (and only if) the addressed location contains a 1. Furthermore, for the instance of the problem that we are considering here, we are promised that one and only one of the locations in the database contains a 1; all other locations contain 0s.

Classically, it would require 3 queries to the database to be certain of the location of the single 1. If, after 3 queries, we had still not found the 1, then we could be certain that it was in the one location in which we had not yet looked. But if we make only 2 queries and do not find the 1 then we have no information about which of the remaining locations may be the correct one; we will have only a 50% chance of choosing correctly after 2 failed queries.

Quantum mechanically the situation is quite different. In fact, a single query to the database allows us to determine, with 100% certainty, the location of the 1. This is accomplished by querying the database with the address qubits in a superposition of all possible addresses, and then decoding the resulting state to extract the needed information.

One version of the quantum algorithm that solves this problem, found by genetic programming and simplified by Herbert J. Bernstein (personal communication), can be expressed as a QGAME program as follows:

```
((HADAMARD 2)
 (HADAMARD 1)
 (U-THETA 0 0.7853981633974483)   ;θ = π/4
 (ORACLE ORACLE-TT 2 1 0)
 (HADAMARD 2)
 (CNOT 2 1)
 (HADAMARD 2)
 (U-THETA 2 1.5707963267948966)   ;θ = π/2
 (U-THETA 1 1.5707963267948966))  ;θ = π/2
```

Figure 3.3 shows a gate array diagram for this same version of Grover's algorithm.

Before running this program we set all qubits to the 0 state. We then run the program and read the answer, which will be the address of the 1

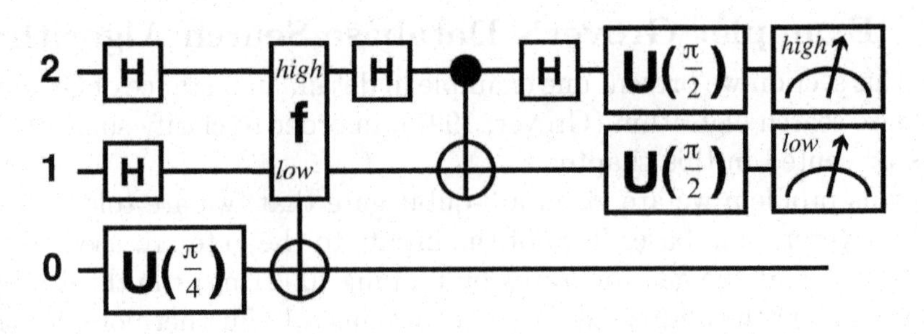

Figure 3.3. A gate array diagram for one version of Grover's database search algorithm for a 4-item database.

in the database, from qubits 2 and 1. Simulation with QGAME confirms that this provides the correct answer in all cases. This simulation can be performed using `TEST-QUANTUM-PROGRAM` with the following inputs:

- `program`: (As listed above).
- `num-qubits`: 3.
- `cases`: (((1000)0)((0100)1)((0010)2)((0001)3)).
- `final-measurement-qubits`: (2 1)
- `threshold`: 0.48.

This call, using the current Lisp implementation of QGAME, produces the following results:

- misses: 0.
- maximum error: $6.661338147750939 \times 10^{-16}$ (zero aside from a tiny round-off error).
- average error: $6.661338147750939 \times 10^{-16}$ (zero aside from a tiny round-off error).
- Maximum expected oracle calls: 1.
- Average expected oracle calls: 1.

Note that the "output" of the database is not even consulted after the database query; instead, the answer is decoded from the states in which the input qubits are left. This highlights a counter-intuitive property of many quantum algorithms, sometimes called the "back action" of unitary gates.

Figures 3.4 through 3.13 illustrate the action of this algorithm via cube diagrams for the single case of a database with the item stored at the address $(0,0)$. Note that measurement of the system's state after the query to the database (as illustrated in Figure 3.8) would produce completely random results; the "decoding" steps in the remainder of the algorithm are necessary to extract the correct answer. Note also that the value of qubit 0, which is nominally the output of the database query, is completely uncertain at the end of the simulation (as illustrated in Figure 3.13).

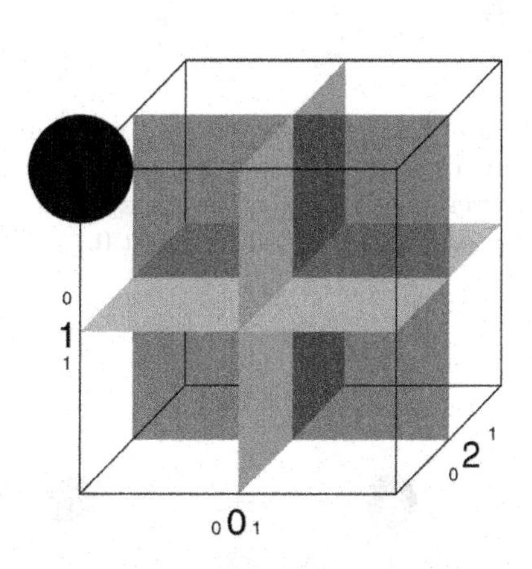

Figure 3.4. A cube diagram of the initial state for Grover's algorithm, as diagrammed in Figure 3.3. All qubits are in the 0 state.

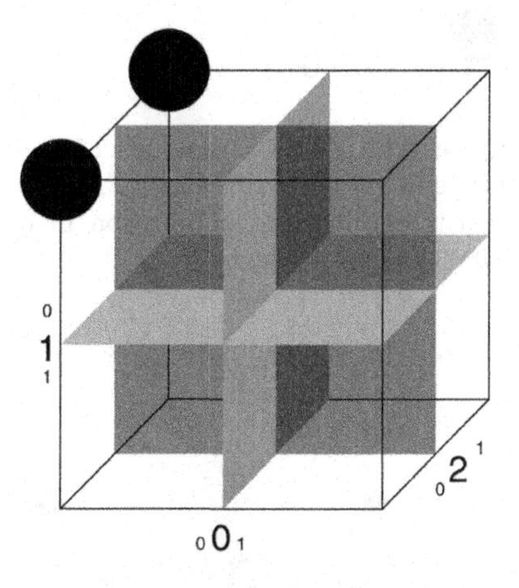

Figure 3.5. A cube diagram of the second state in the execution of Grover's algorithm, after the application of a `HADAMARD` gate to qubit 2.

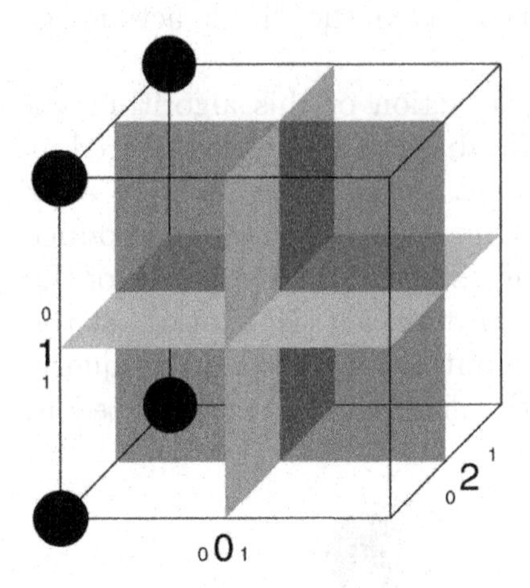

Figure 3.6. A cube diagram of the third state in the execution of Grover's algorithm, after the application of a `HADAMARD` gate to qubit 1.

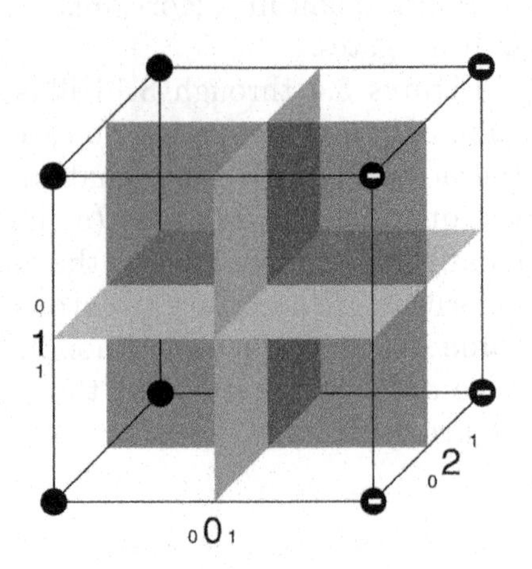

Figure 3.7. A cube diagram of the fourth state in the execution of Grover's algorithm, after the application of a `U-THETA` gate to qubit 0.

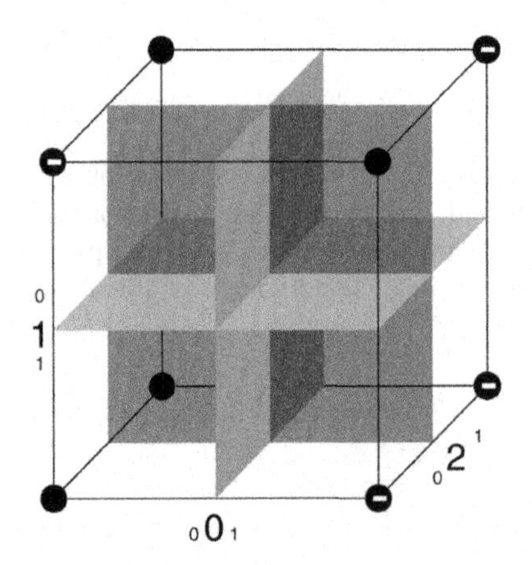

Figure 3.8. A cube diagram of the fifth state in the execution of Grover's algorithm, after the database call. In this example the single 1 in the database is at address (0, 0).

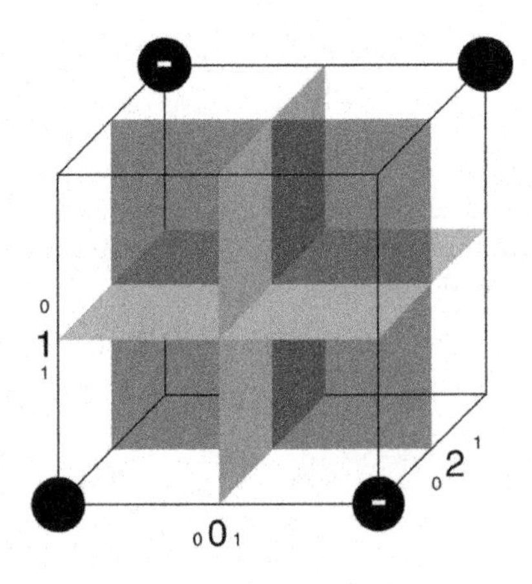

Figure 3.9. A cube diagram of the sixth state in the execution of Grover's algorithm, after the application of another `HADAMARD` gate to qubit 2.

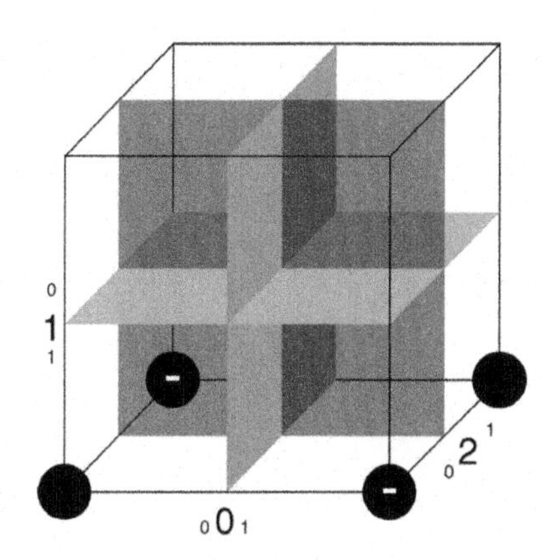

Figure 3.10. A cube diagram of the seventh state in the execution of Grover's algorithm, after the application of a **CNOT** gate with qubit 2 as the control and qubit 1 as the target.

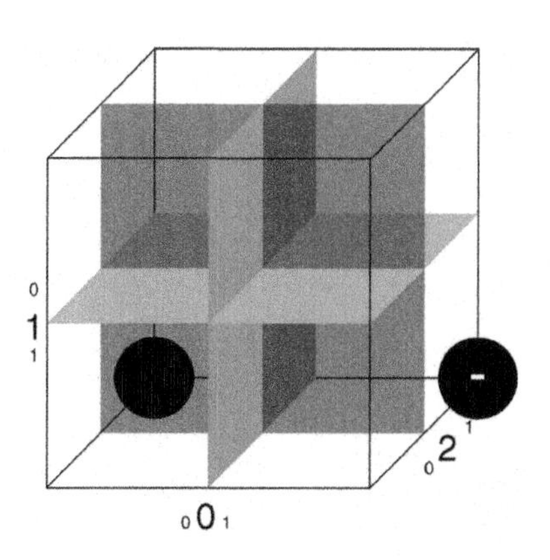

Figure 3.11. A cube diagram of the eighth state in the execution of Grover's algorithm, after the application of another **HADAMARD** gate to qubit 2.

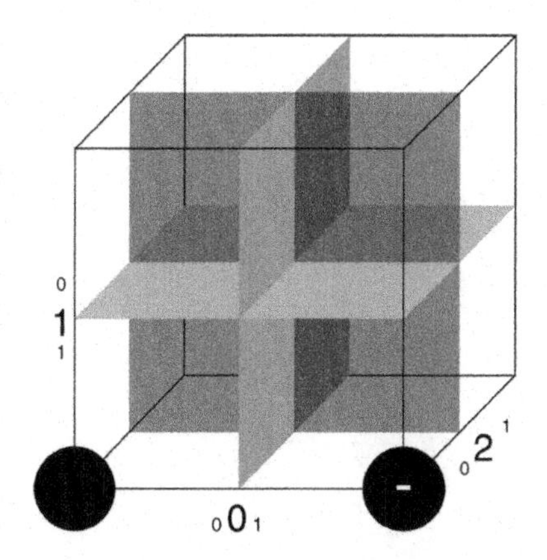

Figure 3.12. A cube diagram of the ninth state in the execution of Grover's algorithm, after the application of a **U-THETA** gate to qubit 2.

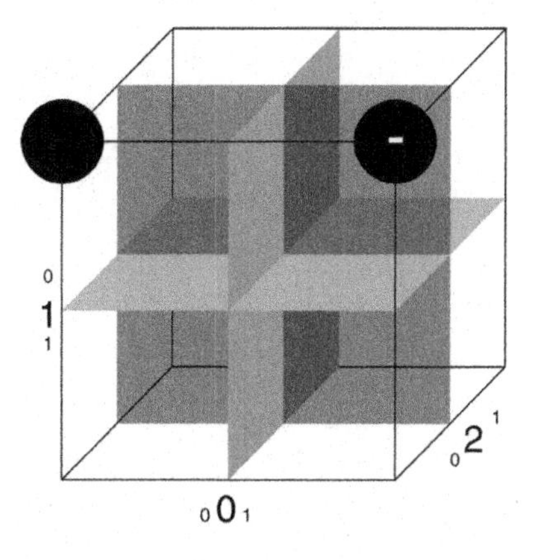

Figure 3.13. A cube diagram of the final state in the execution of Grover's algorithm, after the application of a **U-THETA** gate to qubit 1. The (correct) answer is read from qubits 2 and 1, both of which now have the value 0 with certainty.

Chapter 4

GENETIC AND EVOLUTIONARY COMPUTATION

This chapter introduces genetic and evolutionary computing, focusing on the traditional genetic algorithm. It also discusses, in general terms, the use of parallelism to scale up genetic and evolutionary computation technologies for complex applications, and the applicability of these technologies for various types of problems including those related to quantum computing.

1. What is Genetic and Evolutionary Computation?

The phrase "genetic and evolutionary computation" is used in the literature to describe a wide array of computational enterprises that borrow general principles from genetics and from evolutionary biology. The motivations for these enterprises vary considerably. Some researchers are primarily interested in the processes that underlie biological genetics and evolution, and they use computational models (which may include problem-solving components) as tools to develop, test, and refine biological theory. Others are primarily interested in the problem-solving potential exhibited by evolution and by living systems, and they borrow methods from nature mainly for the sake of engineering more powerful problem-solving systems. And of course many researchers combine both of these motivations, perhaps with others as well.

In this book the focus is on the engineering applications of genetic and evolutionary computation; we seek methods by which the space of quantum algorithms can be explored, and we turn to genetic and evolutionary computation because it provides powerful problem-solving methods that are well suited to this application area. While fidelity to natural genetic

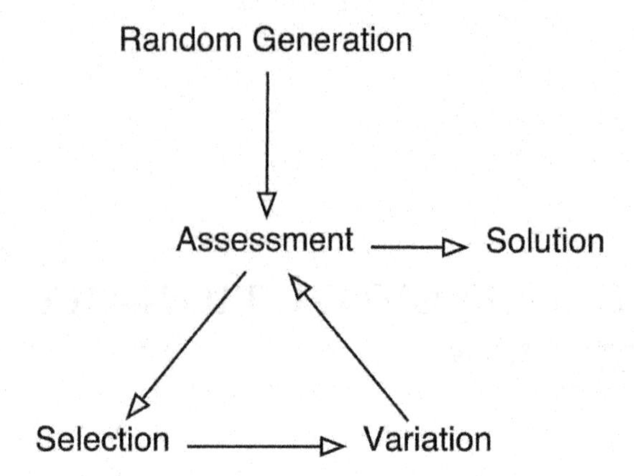

Figure 4.1. A problem-solving strategy common to many forms of genetic and evolutionary computation.

and evolutionary systems is not our primary concern, insights from biology may nonetheless be essential. Nature's "problem-solving methods" are not yet completely understood, and we, as engineers, cannot yet be sure which of nature's methods will serve us best. Indeed, the argument can be made that the cutting edge of practice in genetic and evolutionary computation is moving ever more swiftly toward biology, largely because biological systems still outstrip our technologies in terms of adaptive capacity. For example, recent advances in genetic programming techniques use mechanisms derived from DNA dynamics, learning mechanisms in neural networks, immune systems, regulatory networks, and biological gene expression processes (Spector, 2003). A few of these advances, in particular some of those related to the development and evolution of modular structures, will be discussed in later chapters of this book.

Many genetic and evolutionary computation systems conform to the general structure shown in Figure 4.1. We begin with a population of random individuals. In a problem-solving context an "individual" is usually a candidate solution — something selected from the (normally immense) set of the *kinds* of things that could possibly be solutions. Of course, since the initial population is random it is unlikely that any of these individuals will actually *be* a solution, but these individuals become the starting points for an evolutionary search of the space of candidate solutions.

We then assess the fitness of the individuals in the population. The term "fitness" is used here, as in most genetic and evolutionary computation literature, in a sense different from that which it normally has in biology: it means the value of an individual relative to one's problem-solving goal. The biologist's notion of "reproductive fitness" is useful in genetic and evolutionary computation as well, but it applies here not to the goal-relative "fitness" measure alone, but rather to the combined effects of goal-relative fitness, selection, and variation. Although there are "ecological" and "co-evolutionary" exceptions, fitness is normally assessed for each individual in isolation from the remainder of the population and from any significant external environment.

We use the goal-oriented fitness measure both to drive evolution and to determine success. If, during fitness assessment, a solution to the posed problem is found, then the solution is produced as output and the system halts. Until this happens (or until the user gives up) the system proceeds through a loop of selection, variation, and re-assessment.

The details of this "selection, variation, assessment" loop, and of the representations and algorithms used within it, vary among different forms of genetic and evolutionary computation. For example, in some ("generational") methods the entire population is assessed first and is then subjected to distinct population-wide selection and variation procedures. In other ("steady state") methods single individuals or small groups of individuals are progressed through the entire loop independently. Selection may be based on virtual roulette wheels or on tournaments or on other abstractions. Variation may be asexual (mutation) or sexual (recombination or crossover), and may come in many forms; researchers have experimented with dozens if not hundreds of different mutation and recombination operators.

The extent to which some genetic and evolutionary computation variants might be better than others, in general or for certain sorts of applications, has been a topic of considerable interest in the research community. Many of these questions are addressed within discussions of so-called "No Free Lunch" theorems (for example, Wolpert and Macready, 1997; Droste et al., 1999; Whitley, 1999; Christensen and Oppacher, 2001; Schumacher et al., 2001; Igel and Toussaint, 2003; Woodward and Neil, 2003). Aside from noting that this range of variation exists, along with the associated discussion (and meta-discussion) of the relative merits of the variants, we will not directly address these issues further in this book; we present methods that can be applied to the task of automatic quantum computer programming, and we leave it to the reader to consider or experiment with variations that may produce better results.

2. Genetic Algorithms

One of the most straightforward and widely applied forms of genetic and evolutionary computation is the "genetic algorithm" (GA) as developed by John Holland (Holland, 1992). In the simplest genetic algorithm each individual is a linear chromosome consisting of some fixed number of loci. At each locus of each chromosome is, in the simplest case, a single bit (0 or 1), although sometimes a larger or even unbounded (e.g. continuous) "genetic alphabet" is used. Each chromosome encodes a potential solution to the target problem in some problem-specific way. For example, for one problem each locus of a chromosome might encode a direction to turn in a maze, while for another problem the loci may encode coefficients of a polynomial. Whatever the encoding, the fitness function takes a chromosome as input and produces a fitness value as output.

The traditional genetic algorithm is generational. All chromosomes are assessed for fitness, and then chromosomes are selected, in a fitness-proportionate way, to contribute to the next generation via reproduction and genetic operators.

One popular selection method is "tournament selection." In tournament selection we designate a tournament size T (usually between 2 and 7), and each time that we want to select a chromosome we first randomly select T chromosomes. We then compare the fitness values of the chromosomes and return, as the selection, the one with the best fitness. This method is simple to implement and it allows for selection pressure to be adjusted by changing the tournament size.

The most common genetic operators are point mutations, in which the contents of single loci are probabilistically replaced or perturbed (e.g. by Gaussian noise), and crossovers, in which an offspring is produced by concatenating a left-hand portion of one chromosome with a right-hand portion of another, with the "crossover point" chosen randomly. Variations on these operators abound; for example, in "uniform crossover" each locus is randomly selected from one or the other of two parents. In some schemes the meaning of a locus is de-coupled from the position of the locus on the chromosome, allowing the system to learn appropriate "linkages" between loci, rather than having them determined by a pre-defined sequence. Some schemes have been derived purely on the basis of engineering considerations, while others, for example the use of "diploid" chromosomes, have been derived from biology.

Genetic algorithms have been the subject of intensive study and development and many successful systems, developed according to a wide variety of designs, have been fielded in a wide range of application areas. Good introductory-level treatments of genetic algorithms and their applications include (Goldberg, 1989) and (Mitchell, 1996).

3. Scalability via Parallelism

Because genetic and evolutionary computation systems process populations, and because many of the operations that are performed on the population's individuals are performed independently for each individual (e.g., many forms of fitness assessment), these systems are well suited to parallelization across multiple computer systems. Indeed, loose coupling of multiple sub-populations (often called "demes") with occasional migrations can actually be advantageous to the evolutionary process by slowing the fixation of sub-optimal genetic patterns throughout the system. One can therefore deploy genetic and evolutionary computation systems across large clusters that have moderate or low interconnection bandwidth, thereby reaping gains both in overall computational throughput and in search performance. For this reason genetic and evolutionary computing methods are sometimes referred to in the literature as "embarrassingly parallel."

Parallelization is important for applications to automatic quantum computer programming because these applications often call for quantum computer simulation in the fitness test. As discussed in Chapter 2, the classical simulation of quantum algorithms generally entails exponential inefficiencies, so fitness tests that rely on such simulation will require significant time and/or memory. We cannot fully regain the exponential losses through parallelism (unless we can afford to grow our computer cluster exponentially!), but we can nonetheless expand the range of quantum computation problems that can be addressed by deploying our genetic and evolutionary computation systems across modest-sized computer clusters.

4. Applicability of Genetic and Evolutionary Computation

Genetic and evolutionary computation methods are powerful in part because they require little advance knowledge of the problems being posed or of the structure of possible solutions. Methods with this property are sometimes called "weak methods" in the literature, but this designation actually implies generality, not lack of power.

Genetic and evolutionary computation methods leverage computational resources (CPU cycles) to search vast spaces, combining "blind" exploration (e.g. random initial populations and random variation) with goal-directed guidance (via fitness-based selection, and often via the combinatorics of genetic recombination). As such they are ideal for exploring domains about which we have little prior knowledge. This makes them very powerful tools in the engineer's toolkit.

But by the same token genetic and evolutionary computation technologies, like all so-called weak methods, will generally underperform specialized methods that are based on a deep understanding of a particular problem area's search space. When one knows a domain sufficiently well one can often develop problem-solving methods that are considerably more efficient than random variation and selection.

Owing to our current relative ignorance about the nature of quantum algorithms, about the principles of quantum software engineering, and about quantum complexity theory, the applicability of genetic and evolutionary methods to automatic quantum computer programming appears to be strong at present. It may weaken, however, as the fields of quantum computing and quantum information theory mature.

Chapter 5

GENETIC PROGRAMMING

This book is concerned with automatic quantum computer *programming* by means of genetic and evolutionary computation. In Chapter 4 we described genetic and evolutionary computation methods in general; here we narrow the focus to the form of genetic and evolutionary computation most directly concerned with the discovery of programs, namely "genetic programming." We provide a concise introduction to the basic concepts of genetic programming, a detailed example, and a discussion of the steps that one must generally take to obtain and understand useful results from a genetic programming system. The techniques described in this chapter are not specific to *quantum* computing; we will narrow the focus further to genetic programming *for quantum computers* in Chapter 7, following the description, in Chapter 6, of advanced genetic programming techniques that are particularly useful for evolving quantum programs.

1. Programming by Genetic Algorithm

A "genetic programming" system is a genetic algorithm in which the chromosomes are executable computer programs. There is no sharp line between "executable computer programs" and other chromosomal encodings at a fundamental level of analysis, since the elements of any encoding could be considered "commands" in some language for which the chromosome-decoder is the "compiler" or "interpreter." But in practice so-called genetic programming systems tend to differ from other genetic algorithms in several ways. Although there are exceptions, genetic programming systems tend to use chromosome encodings that are similar in syntax and semantics to existing programming languages. They tend to allow chromosomes to vary in length (as computer programs normally do) and to incorporate hierarchical, compositional structures. And they often allow one to combine the use of special-purpose, problem-specific instructions or variables with general-purpose, commonly-used instructions (for example, $+$ and $\times$).

As a result, a genetic programming system is not only a problem-solving system but also an automatic programming system. We provide the system with information that describes what we want a program to do (primarily via the fitness function), and a successful run of the system produces (via evolution) a program that meets the desired specification. We must also provide values for a few other parameters (such as the set of instructions that *may* be used, population sizes, and genetic operator rates), but most of this is straightforward and requires little expertise about the problem we are using the system to solve.

In other words, what we provide to a genetic programming system is mostly just a specification of the behavior of the program that we seek. A successful run of the system produces, automatically, a program that exhibits the desired behavior. So genetic programming is an automatic programming technology that produces programs via genetic and evolutionary computation.

As John Koza has argued forcefully in several of his books, automatic programming is an extremely general capability that can be applied in almost every conceivable area of science and technology (Koza, 1992; Koza, 1994; Koza et al., 1999; Koza et al., 2003). Genetic programming in particular has been applied to a wide range of problems, including many in science and technology and even several in the arts (for example, Spector and Alpern, 1994; Polito et al., 1997).

Beyond its applicability to "external" problems, automatic programming (and thereby genetic programming) also opens up new approaches to the study of fundamental questions in computer science itself. This is because many fundamental questions in computer science are about whether or not there exist computer programs having particular properties. Many of these questions can be approached using analytical techniques; for example, mathematical proofs are often employed to demonstrate negative results, proving that *no* program can possibly have some particular set of properties. But many other questions are best approached via algorithm design — researchers attempt to find programs that have the properties in question, usually relying on their own experience and ingenuity to do so.

Automatic programming technologies open the door to new approaches to such questions, allowing us to use the computer itself to search the space of computer programs and thereby to expand the frontiers of theoretical computer science. They allow us to do "computer science by automatic programming" — and if our automatic programming technology is genetic programming, then we can do "computer science by genetic programming." This is in fact the goal of the present book, which seeks more specifically to describe some of the ways in which one can do "quantum computer science by genetic programming."

The most widely used genetic programming techniques are documented in Koza's books (Koza, 1992; Koza, 1994; Koza et al., 1999; Koza et al., 2003), although some readers may prefer the more concise introduction in (Banzhaf et al., 1998), which also includes a survey of alternative approaches such as "machine code" genetic programming. Innovations in genetic programming technique are regularly reported at several international conferences with published proceedings, most notably the *Genetic and Evolutionary Computation Conference* (GECCO), which combines the previously existing *Genetic Programming Conference* and the *International Conference on Genetic Algorithms.* Important journals in the field include *Genetic Programming and Evolvable Machines* and *Evolutionary Computation.* Advances in genetic programming technique have also been documented in several edited books (Kinnear, Jr., 1994a; Angeline and Kinnear, Jr., 1996; Spector et al., 1999c; Riolo and Worzel, 2003). A searchable, online bibliography on genetic programming is also available.[1]

2. Traditional Program Representations

It is a simple matter to suggest the use of executable computer programs as chromosomes in a genetic algorithm, but it is more difficult to devise detailed schemes for program representations and genetic operators that allow such a genetic algorithm to perform well. The most popular scheme for genetic programming, which is now often called "traditional" genetic programming, "standard" genetic programming, or "tree-based" genetic programming (for reasons that will be made clear below), was developed and popularized primarily by John Koza, although similar ideas were also presented early by others (Cramer, 1985; Koza, 1992).

One obvious concern in using programs as chromosomes is that programs must generally conform to a particular syntax — that of the programming language in which they are expressed — in order to be meaningful at all. The misplacement of a single character in a program in most languages is very often "fatal," in that the program will fail even to compile or will cause an interpreter to halt abnormally in an error

[1] `http://liinwww.ira.uka.de/bibliography/Ai/genetic.programming.html`

condition. The genetic operators in a genetic algorithm are generally "blind" — they "slice and dice" chromosomes without any cognizance of chromosomal syntax. How then are we to ensure that the "offspring" of syntactically valid programs, as produced by genetic operators, are always (or at least often) themselves syntactically valid?

The traditional approach to this concern is to use a syntactically minimal programming language as the language for the evolving (chromosomal) programs. Although there are many candidates for such a language (and several have been explored), the one traditionally used is a tiny subset of Lisp.

Lisp is a language with a long and important history in computer science (see, for example, McCarthy et al., 1966, Steele Jr., 1984, and Graham, 1994), but few of its significant features are exploited in traditional genetic programming. For the most part, all that is used is the basic syntax for programs (which, in Lisp, are also data). The subset of Lisp syntax used in traditional genetic programming is usually just:

$$program \quad ::= \quad terminal \ \mid \ (\ function_n \quad program^n \ )$$

In other words:

- A "terminal" is a program; terminals are often constants (like "5" or "3.14" or "TRUE") although they may also be variables or zero-argument functions.

- A parenthesized sequence of an n-ary function followed by n additional programs is a program. This is a *prefix* notation for a function call with n arguments.

A simple example is an expression like:

```
(+ (* X Y)
   (+ 4 (- Z 23)))
```

This expression is interpreted "functionally," with each program returning a value to the enclosing context, usually to be passed as an argument to the function heading the enclosing expression. Overall this example returns the sum of two values, the first of which is the product of `X` and `Y` (which are presumably variables containing numbers), and the second of which is the sum 4 and `Z` minus 23.

Functional interpretation (that is, interpretation in which each function call's primary job is to return a value to the enclosing context) is not mandatory, however, and many genetic programming applications

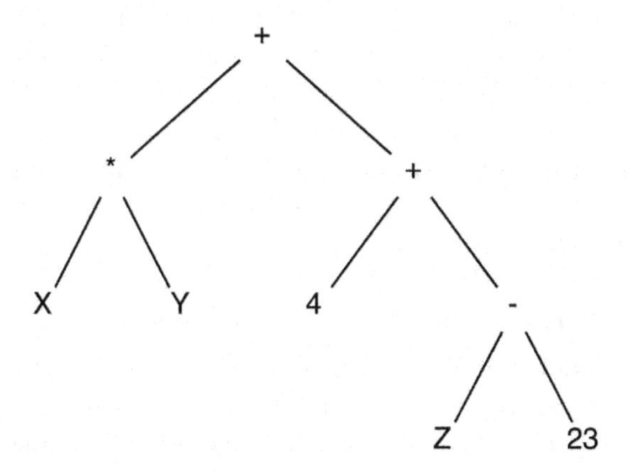

Figure 5.1. A tree graph of the arithmetic Lisp expression given in the text.

have utilized functions that act primarily by side-effect on external data structures. For example, in the classic "artificial ant" problem (Koza, 1992) the terminals (`LEFT`, `RIGHT`, and `MOVE`) are interpreted as commands to a simulated ant moving on a grid containing simulated food. They return no values, and the functions that can appear as the first items of parenthesized expressions "expect" no arguments; they merely serve to sequence the calls to the side-effecting terminals (in some cases conditionally).

The essential feature of this program representation with respect to genetic programming is syntactic uniformity — any sub-program can be substituted for any other sub-program within any program, and the result will necessarily be syntactically well formed. It is therefore easy to devise genetic operators that operate "blindly" on programs but nonetheless always produce syntactically valid results. These representations are often called "tree-based" because they can be presented graphically using tree structures as in Figure 5.1, which shows the tree form of the arithmetic expression given above.

In traditional genetic programming all of the constant terminals used for a particular run must be of the same data type. The functions used in the run must all return values of this same type, and must take arguments only of this type. These restrictions prevent type incompatibility errors, but they are inconvenient; several ways to relax these restrictions are discussed in Chapter 6.

Additional steps must often be taken to ensure that arbitrary programs are also *semantically* valid — that is, that they will always execute without error, producing interpretable (even if incorrect) results. For example, one must sometimes engineer special return values for "pathological" calls, such as division by zero. To handle division by zero, one usually replaces the division function with a "protected division" function that returns an arbitrary value (for example, zero or one) when it receives zero as its second argument (Koza, 1992). Similar strategies can be employed to produce "protected" versions of other functions with pathological special cases.

3. Traditional Genetic Operators

The most common forms of genetic programming mutation involve the replacement of an arbitrarily chosen subprogram of with a newly generated random subprogram. For example, consider the following program:

```
(+ (* X Y)
   (+ 4 (- Z 23)))
```

If we wish to mutate this program we first select a random subprogram for replacement, as follows:

```
(+ [(* X Y)]
   (+ 4 (- Z 23)))
```

We then generate a new random subprogram and insert it in place of the selected subprogram:

```
(+ (- (+ 2 2) Z)
   (+ 4 (- Z 23)))
```

There is plenty of room for variation of this basic scheme for mutation and many variants have been explored. For example, it is common to bias the selection of subprograms in favor of entire function calls (rather than terminals) (Koza, 1992), and one can vary the ways in which random subprograms are generated (for example, to limit their length). One variation that may help to prevent run-away program growth ("bloat") forces replacement subprograms to be similar in size to the subprograms that they replace; this is called "size fair mutation" (Langdon et al., 1999; Crawford-Marks and Spector, 2002). A survey of published mutation operators appears in (Langdon, 1998).

Program crossover (recombination) is usually accomplished in a similar way, via the swapping of subprograms. Given two parent programs (which may be selected from the population on the basis of fitness tournaments or other fitness-sensitive selection methods), we select random subprograms in each:

```
Parent 1:   (+ [(* X Y)]
               (+ 4 (- Z 23)))
Parent 2:   (- (* 17 (+ 2 X))
               (* [(- (* 2 Z) 1)]
                  (+ 14 (/ Y X))))
```

We then swap the subprograms to produce two potential child programs:

```
Child 1:   (+ [(- (* 2 Z) 1)]
              (+ 4 (- Z 23)))
Child 2:   (- (* 17 (+ 2 X))
              (* [(* X Y)]
                 (+ 14 (/ Y X))))
```

Again, many variations have been proposed and tested, including variations intended to increase the chances that children of fit parents will themselves be fit, variations intended to increase or decrease the "exploratory power" of recombination, and variations intended to control the size and shape statistics of evolving populations.

The rates at which these genetic operators are applied — that is, the proportions of a generation produced by mutation, crossover, and exact reproduction — have also been the subject of many studies (for example, Luke and Spector, 1998).

4. Example: Symbolic Regression

This section provides a brief example of traditional genetic programming applied to a "symbolic regression" problem. The task in symbolic regression (Koza, 1992) is to find an equation that fits a provided set of data. Many other types of regression analysis require the user to specify the form of the solution (for example, linear or quadratic) in advance; by contrast, in symbolic regression we have no *a priori* knowledge of the form of the solution and we expect the genetic programming system to find both the form and the details of the solution equation.

The example that we consider here is a 2-dimensional symbolic regression problem, in which we are given a set of (x, y) pairs and the task of producing a program that takes an x value as input and produces the appropriate y value as output. For the example here we generated the data from the function $y = x^3 - 0.2$, using 20 x values evenly distributed between zero and one. Of course the system is given only the data and not the function that generated the data; the task of the system is to re-discover the generating function.

Table 5.1. Parameters for the example run of traditional genetic programming on a symbolic regression problem. The "%" function is a protected division function that returns 1 if its second argument is 0. Detailed explanations of these parameters can be found in (Koza, 1992).

Target function	$y = x^3 - 0.2$
Function set	$\{+, -, *, \%\}$
Terminal set	$\{x, 0.1\}$
Maximum number of Generations	51
Size of Population	1000
Maximum depth of new individuals	6
Maximum depth of new subtrees for mutants	4
Maximum depth of individuals after crossover	17
Fitness-proportionate reproduction fraction	0.1
Crossover at any point fraction	0.3
Crossover at function points fraction	0.5
Selection method	fitness-proportionate
Generation method	ramped-half-and-half
Randomizer seed	1.2

Koza's "Little Lisp" demonstration genetic programming code[2] was used for the run described below, with the parameters shown in Table 5.1. For each fitness test the program under consideration was evaluated for each of the 20 x values. Each such "fitness case" produced an error value, calculated as the absolute value of the difference between the y value produced by the program and the y value corresponding to the input x value in the data set. The sum of the errors over the 20 fitness cases was taken to be the overall "fitness" of the program, with lower fitness values indicating better programs. The fitness value for a perfect program, using this scheme, is zero.

In the initial, randomly generated population (generation 0), the program with the best (lowest) fitness was as follows:

```
(- (% (* 0.1
         (* X X))
      (- (% 0.1 0.1)
         (* X X)))
   0.1)
```

[2]http://www.genetic-programming.org/gplittlelisp.html

This program has a fitness value (total error) of about 2.22127 and is shown graphed against the target function ($y = x^3 - 0.2$) in Figure 5.2. The best fitness in the population generally improved each generation, with the best program in generation 5 being the following:

```
(- (* (* (% X 0.1)
         (* 0.1 X))
      (- X
         (% 0.1 X)))
   0.1)
```

This program has a fitness of 1.05 and and is shown graphed against the target function in Figure 5.3. By generation 12 a considerably better program was found, with a fitness value of 0.56125:

```
(+ (- (- 0.1
         (- 0.1
            (- (* X X)
               (+ 0.1
                  (- 0.1
                     (* 0.1
                        0.1))))))
      (* X
         (* (% 0.1
               (% (* (* (- 0.1 0.1)
                        (+ X
                           (- 0.1 0.1)))
                     X)
                  (+ X (+ (- X 0.1)
                          (* X X)))))
            (+ 0.1 (+ 0.1 X)))))
   (* X X))
```

This program is shown graphed against the target function in Figure 5.4. Although this program is large, some of the code that it contains is "junk" because, for example, it produces a result that is later multiplied by zero. Issues related to such non-functional code and its possible contributions to code "bloat" and evolutionary progress have been discussed extensively in the literature; see for example (Luke, 2000).

In this run a perfect solution (fitness 0) was found in generation 22 (Figure 5.5), in the following form:

```
(- (- (* X (* X X)) 0.1) 0.1)
```

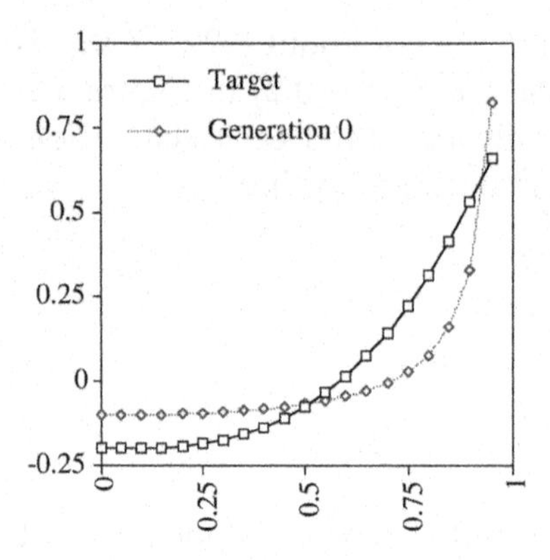

Figure 5.2. The performance of the best program of generation 0, plotted against that of the target function, in an application of standard genetic programming to a symbolic regression problem.

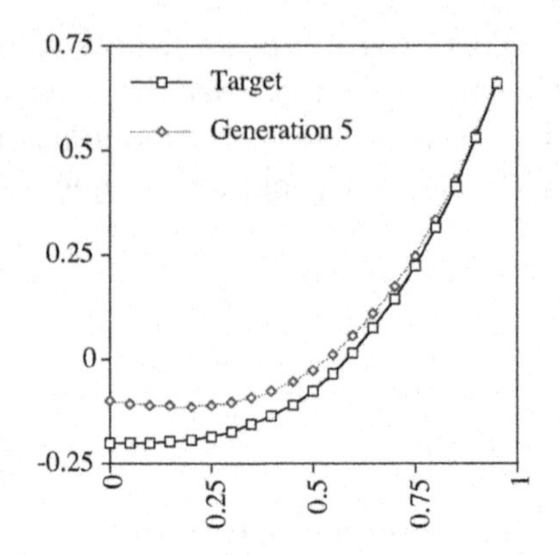

Figure 5.3. The performance of the best program of generation 5.

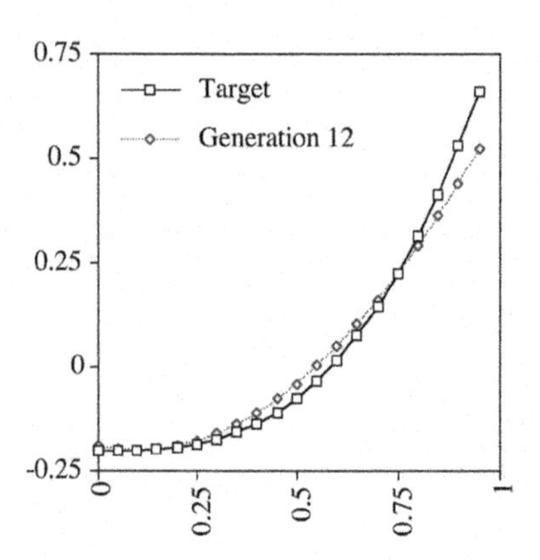

Figure 5.4. The performance of the best program of generation 12.

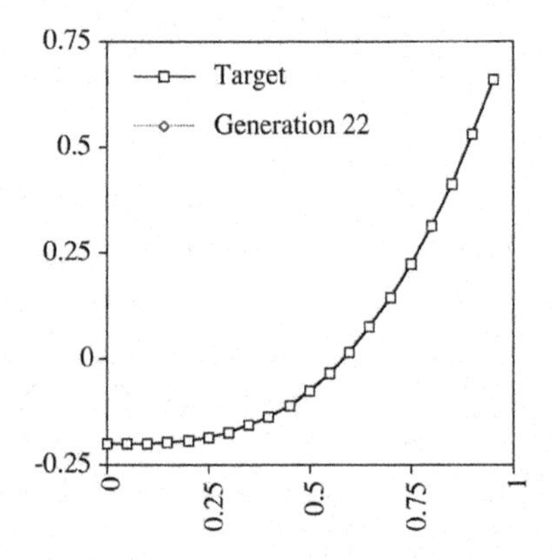

Figure 5.5. At generation 22, a perfect match to the target function is found.

5. Obtaining Genetic Programming Results

Genetic programming is a general technique that can be applied, without substantial re-engineering, to a wide array of problems. The preparatory steps that one must complete to apply the technique to a new problem include the selection and/or definition of appropriate functions and terminals out of which programs will be constructed, the definition of a problem-specific fitness function, and the setting of other parameters such as population size and mutation and crossover rates. These steps are described in detail by John Koza (Koza, 1992), who also makes the case that one can often obtain good results by making straightforward or standardized choices in each of the preparatory steps. Although solutions may emerge more rapidly or more reliably with a carefully circumscribed function set, or with a refined fitness function, or with a mutation rate that has somehow been adjusted to suit a particular problem, etc., one often finds that the "obvious" or standard choices nonetheless suffice to solve the problems in which one is interested.

Nonetheless, in some cases — particularly when attempting to solve difficult real-world problems — it may be necessary to apply more art than science to genetic programming's preparatory steps. If a problem is resisting solution, for example, then one might want to use a larger population. But larger populations require more processing time, which may make it difficult to run the system for a sufficiently large number of generations. As a consequence one might try running with various population sizes for a small number of generations to get a sense of the rates of progress at each setting, and one might follow up this exploration with an extended run at a particular population size. Similarly, one might notice that the system tends to get "stuck" short of a solution, after which the diversity of the population plummets.[3] In this case one might experiment with different mutation or crossover rates or alternative fitness functions. One may gain other insights, and be led to experiment with other parameters, by watching average program sizes over the course of a run. In short, while standard choices for many parameters may perform reasonably well for a wide range of problems, progress on difficult real-world problems sometimes, nonetheless, demands experimentation and tuning.

Once a solution has been found, it may require further work to *understand* the solution that has evolved. Genetic programming may produce programs that solve problems by means of novel principles, and the essential features of the evolved solutions may be buried in large volumes of irrelevant or non-functional code. A variety of approaches has been applied to this problem of analysis. In some cases it may be relatively easy to edit out non-functional code, such as expressions that produce values that are later multiplied by zero, using knowledge about the func-

[3]Several measures of population diversity have been developed in the genetic programming literature; see for example (Burke et al., 2002a) and (Burke et al., 2002b).

tion and terminal sets and the ways that they interact. In other cases it may be more helpful to use a second phase of genetic programming to minimize the size of the result. For example, one can define the fitness of any genuine solution to be the size of the solution program, and the fitness of any non-solution to be larger than the largest permissible size. One can then conduct a run of genetic programming with an initial population consisting only of previously found (but probably large) solutions, and use the run to minimize solution length. In many cases, however, the only path to understanding a solution produced by genetic programming is to trace carefully the execution of the solution program.

Chapter 6

EVOLUTION OF COMPLEX PROGRAMS

Traditional genetic programming, as described in Chapter 5, is a powerful problem-solving tool but it nonetheless has several limitations. Some of these limitations prevent the successful application of the technique to large-scale, difficult problems such as the automatic quantum computer programming problems discussed in this book. Fortunately, however, an active international community of researchers has enhanced the technique in ways that extend its power significantly; in this chapter several such enhancements are presented, with a focus on those that find direct application in automatic quantum computer programming.

More specifically, this chapter describes some of the ways in which genetic programming techniques can be used to evolve programs that include multiple data types, modules, and developmental components. Although these capabilities were developed for problems unrelated to quantum computing, several of them are nonetheless particularly useful for the evolution of quantum programs. The author's Push programming language for genetic and evolutionary computation, which provides some of the desired advanced capabilities in unusually simple ways, and the author's PushGP genetic programming system, which evolves Push programs, are described in detail. These technologies, while not themselves specifically oriented toward quantum computing, underlie the techniques for automatic quantum computer programming described in Chapter 7, which are in turn used for the production of the results documented in Chapter 8. The chapter concludes with a brief description of self-adaptive "autoconstructive evolution" techniques that are enabled by Push.

1. Types, Modules, and Development

In many application areas it is natural to use several data types. In automatic quantum computer programming, for example, it is natural to use integers (for qubit indices), floating-point numbers (for parameters to quantum gates such as U_θ), and possibly unitary matrices containing complex numbers (for the expression of novel quantum gates). For some problems it may be natural to use even more types, for example arrays of classical (non-quantum) bits, genuine ratios (such as $\frac{2}{3}$, as opposed to 0.666...), etc. In this context, one of the most glaring limitations of the traditional genetic programming technique is the requirement that evolved programs can manipulate values only of a single data type.

One can sometimes work around the single-type limitation of the standard technique by considering all of the required values to be members of a "union" of several data types, and by ensuring that all of the functions in the function set can handle all possible members of this union type in all argument positions. But this is an awkward maneuver that becomes impractical for problems that call for many data types.

To address this need, David Montana has developed an extension to traditional genetic programming called "strongly typed genetic programming." In strongly typed genetic programming one annotates each terminal and each function with type information (Montana, 1993). New procedures, which are sensitive to this type information, are used for the generation of random programs and for genetic operators such as mutation and crossover. As long as these operations all respect the type requirements of the functions and terminals used by the system, the remainder of the genetic programming process can proceed unchanged.

Strongly typed genetic programming allows for the evolution of programs that manipulate multiple types, although it presumably also has impacts on evolutionary dynamics. For example, because strongly typed crossover can swap subprograms only if they return the same types, there will generally be many fewer crossover options for a pair of strongly typed programs than there would be for a pair of untyped programs of similar sizes. It is not clear if these impacts are generally beneficial, detrimental, or neutral, but it is clear in any event that Montana's technique allows genetic programming to be applied to a wider range of problems. Practitioners have found it to be useful in many situations and developers have incorporated it into many genetic programming systems.

Another limitation of the standard technique that should be clear to any student of programming languages is the lack of facilities for the expression of subroutines or other modular code structures. For automatic quantum computer programming modular structures may be particularly helpful because, for example, one often wishes to perform an identical transformation on each of several qubits in a quantum register. Some forms of modular code structure, based on conditionals ("IF THEN" structures) and iterative loops ("DO UNTIL" structures), can be incorporated into genetic programming in straightforward ways (Koza, 1992), but these fall short of the code structuring facilities provided in even the most rudimentary programming languages designed for human use. For any complex program it is usually advantageous to design blocks of code that can be expressed once and then reused multiple times with different inputs over the course of a single program execution.

A variety of schemes have been proposed for the incorporation of modules (sometimes also called subroutines, defined functions, automatically defined functions, automatically defined macros, or products of encapsulation) in the programs that are manipulated and produced by genetic programming (see for example Koza, 1990; Koza, 1992; Angeline and Pollack, 1993; Kinnear, Jr., 1994b; Spector, 1996; Racine et al., 1998; and Roberts et al., 2001). The most popular of these is probably the "Automatically Defined Function" (ADF) scheme presented in detail by Koza in his first and second genetic programming books (Koza, 1992; Koza, 1994). In this scheme the structure of the programs in the population is restricted to a pre-specified modular architecture, with some fixed number of function definitions (each of which takes some fixed number of arguments) and a "result-producing branch." One also specifies, in advance, which automatically defined functions can call which other automatically defined functions. Program generation and manipulation procedures (for example, the procedures for mutation and crossover) are all refined to respect the restrictions of the specified modular architecture.

Koza showed that the use of ADFs allows genetic programming to exploit regularities inherent in many problems and thereby to scale up to substantially larger problem instances. He also showed that the use of ADFs usually allows solutions to be found more quickly and that the solutions so found are usually more parsimonious than those found by the traditional genetic programming technique. In subsequent work he showed how one can add another "architecture altering" layer to the genetic programming process to allow ADF architecture to evolve during a run (Koza et al., 1999).

The final limitation of traditional genetic programming to be considered here concerns the evolution of programs or other executable structures that do not easily map to traditional Lisp-derived program representation. For example, a neural network is in some sense a program, but many neural network architectures ("recurrent" architectures) allow

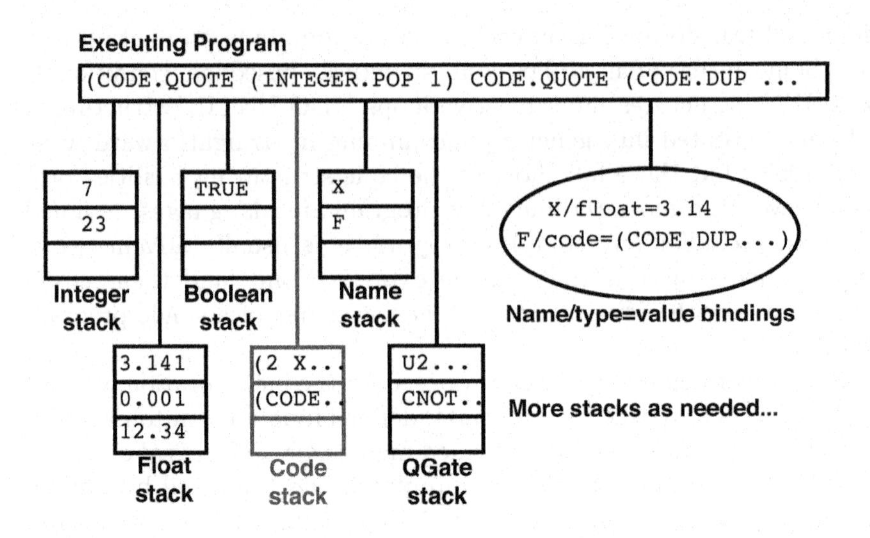

Figure 6.1. A schematic view of a Push interpreter.

for loops that do not map nicely to the tree-like structures of traditional genetic programming. Of course, one can evolve neural networks using other forms of genetic and evolutionary computation; for example, one can use neuron connectivity matrices as chromosomes and evolve networks using traditional genetic algorithm techniques. But several features of the genetic programming paradigm are useful for the evolution of neural networks, including support for variable-length representations and the forms of modular reuse described in the previous paragraphs.

To address this need several researchers have extended the genetic programming technique with "developmental" features. Although there are variations, the basic move in most developmental approaches is to retain traditional program representations but to drop the notion that an evolved program *itself* solves to the problem under consideration. Rather, the evolved (chromosomal) program, when executed, produces (or develops into) *another* program that actually solves the problem. In some cases the structure produced by the execution of the chromosomal program is not a "program" in the traditional sense but instead some other type of "executable object" which is then "run" to determine the behavior and fitness of the individual. For example, in some of John Koza's work the execution of the chromosomal program produces specifications for electrical circuits or control systems (Koza et al., 1999; Koza et al., 2003). In other cases the product of development may be a program in the traditional sense but the chromosome may *not*; for example, in *grammatical evolution* the chromosome is a string of bits or integers which is transformed into a program, during development, by a process that involves indexing into a grammar of a conventional programming language (O'Neill and Ryan, 2003). In yet other cases there may be no clear distinction between the developmental and execution phases; for example, in the *ontogenetic programming* framework (Spector and Stoffel, 1996b; Spector and Stoffel, 1996a), and in the PushGP system described in Section 6.4, programs can modify (develop) their own code *as they run.* While doing so they may also be constructing secondary programs or executable structures which are themselves the solutions to the problems under consideration — if so then the systems of which they are elements are "developmental" in two distinct senses.

Most developmental process must be conducted as a part of every fitness test, although in some cases it is possible to assess a program on multiple "fitness cases" (input sets) after a single developmental phase. Development can be accomplished by various means, with the most common strategy being to begin each developmental phase with a minimal "embryo," to which later function calls add components. The secondary, developed "program" may take various forms, ranging from neural networks (Gruau, 1994) to electrical circuits, control system specifications and even metabolic pathways (Koza et al., 2003). And as we will see in Chapter 7, similar techniques are also useful for the evolution of quantum computer algorithms.

2. The Push Programming Language

Push is a programming language designed specifically for use in genetic and evolutionary computation systems that evolve programs, as the language in which evolving programs are expressed (Spector, 2001; Spector and Robinson, 2002a; Spector et al., 2003b). Push has an unusually simple syntax, which facilitates the implementation (or evolution) of mutation and recombination operators that generate and manipulate programs. Despite this simple syntax, Push provides more expressive power than most other program representations that are used for program evolution. This expressive power allows Push-based genetic and evolutionary computation systems to provide many of the advanced capabilities described in Section 6.1 (along with others) with less system complexity or user configuration.

Push programs can process multiple data types without the syntax restrictions that usually accompany this capability, and they can express and make use of arbitrary control structures (e.g. recursive subroutines and macros) through the explicit manipulation of their own code (via a `CODE` data type). This allows Push to support the automatic evolution of modular program architectures in a particularly simple way, even when it is employed in an otherwise ordinary, ADF-free genetic programming system (such as PushGP; see Section 6.4). Push can also support entirely new evolutionary computation paradigms such as "autoconstructive evolution," in which genetic operators and other components of the evolutionary system themselves evolve (as in the Pushpop and SWARMEVOLVE 2.0 systems; see Section 6.5).

Push achieves its combination of syntactic simplicity and semantic power through the use of a stack-based execution architecture that includes a stack for each data type. This architecture extends the single-type architecture used in previous work on "stack-based genetic programming" (Perkis, 1994; Stoffel and Spector, 1996; Tchernev, 1998).

A diagram of the Push execution architecture is shown in Figure 6.1. The `CODE` data type, which has its own stack and an associated set of code-manipulation instructions, provides many of the more interesting features of the language. Push instructions, like instructions in all stack-based languages, take any arguments that they require and leave any results that they produce on data stacks.

To provide for "stack safe" execution of arbitrary code Push adopts the convention, used widely in stack-based genetic programming, that instructions requiring arguments that are not available (because there are too few values on the relevant stacks) become `NOOP`s; that is, they do nothing. Because Push's stacks are typed, instructions will always receive arguments and produce results of the appropriate types (if they do anything at all), regardless of the contexts in which they occur.

The syntax of Push is simply this:

$$program \;::=\; instruction \;\mid\; literal \;\mid\; (\; program* \;)$$

In other words:

- An instruction is a Push program.
- A literal is a Push program.
- A parenthesized sequence of zero or more Push programs is a Push program.

Parenthesized sequences are also referred to as "lists," and Push programs can in fact be treated as list data structures. Literals are constants such as "`3`" (an integer constant), "`3.14`" (a floating point number constant), and "`TRUE`" (a Boolean constant). Instruction names generally start with the name of the type that they primarily manipulate, followed by a "."; for example, INTEGER.+ is the instruction for adding two integers, and BOOLEAN.DUP is the instruction for duplicating the value on the top of the Boolean stack.

Execution of a Push program involves the recursive application of the following procedure:

```
To execute program P:
    If P is a single instruction then execute it.
    Else if P is a literal then push it onto the
         appropriate stack.
    Else P must be a list; sequentially execute
         each of the Push programs in P.
```

A top-level call to the interpreter can be provided with a list of literals to be pushed onto the appropriate stacks before the program is executed. In addition, the program passed to the top-level call will itself be pushed onto the CODE stack before execution; this convention simplifies the expression of some recursive programs (see below for an example).

The CODE.QUOTE instruction is an exception to the execution procedure given above. Execution of CODE.QUOTE has no immediate effect, aside from changing the state of the interpreter such that the subsequent piece of code considered for execution will not in fact be executed — it will instead be pushed onto the CODE stack. This provides a convenient way to get specific pieces of code onto the CODE stack, where they may be manipulated and/or executed by later instructions.

The NAME data type provides for symbolic variable names and associated binding spaces via GET and SET instructions that are defined for all types. Any identifiers that do not represent known Push instructions or literals of other types (such as TRUE and FALSE) are recognized as NAMEs, and are pushed onto the NAME stack when executed.[1] CODE, like any other type, has a binding space; this means that NAMEs can be used to name subroutines (or pieces of code for any other purpose) in the same way that they can be used to implement variables of other types.

A Push interpreter contains a random code generator that can be used to produce random programs or program fragments. This can be called from outside the interpreter (for example to create or mutate programs in a genetic programming system) or from a standard CODE.RAND

[1]Some implementations of Push may require NAMEs to be distinguished in other ways as well, for example by beginning with a special character such as "_". The NAME type is not used in the examples in Chapter 8, but it is described here both for completeness and to simplify some of the examples in this chapter.

instruction (which is analogous to RAND instructions available for other types). Several algorithms for the generation of random code have been described in the genetic programming literature. Random code generation is less complicated for Push programs than it is for Lisp-style code trees, since in Push one doesn't have to worry about function "arity" or about function versus argument positions when generating code. So it is easier, for example, to generate programs with predictable size and shape distributions. The standard Push random code generation algorithm is shown in Figure 6.2. It produces a uniform distribution of sizes and what seems to be a reasonable distribution of shapes, in a reasonable amount of time. An "ephemeral random constant" mechanism, similar to that employed in traditional genetic programming (Koza, 1992), allows randomly-generated code to include newly-generated literals of various types.

Execution safety is an essential feature of Push, in the sense that any syntactically correct program should execute without crashing or signaling an interrupt to the calling program. This is because Push is intended for use in genetic and evolutionary computing systems, which often require that bizarre programs (for example those that result from random mutations) be interpreted without interrupting the evolutionary process. The "stack safety" convention described above (that is, the convention that any instruction that finds insufficient arguments on the stacks acts as a NOOP) is one component of this feature. In addition, all instructions are written in ways that are internally safe; they have well defined behavior for all predictable inputs, and they typically act as NOOPs in predictable "exceptional" situations (like division by zero).

Additional safety concerns derive from the availability of explicit code manipulation and recursive execution instructions, which can in some cases produce exponential code growth or non-terminating programs. In response to these concerns Push interpreters enforce two limits:

EVALPUSH-LIMIT: This is the maximum allowed number of "executions" in a single top-level call to the interpreter. The execution of a single Push instruction counts as one execution, as does the processing of a single literal, as does the descent into one layer of parentheses (that is, the processing of the "(" counts as one execution). When this limit is exceeded the interpreter terminates execution immediately, leaving its stacks in the states they were in prior to termination (so they may still be examined by a calling program). Whether or not this counts as an "abnormal" termination is up to the calling program.

MAX-POINTS-IN-PROGRAM: This is the maximum size of an item on the CODE stack, expressed as a number of points. A point is an instruction, a literal, or a pair of parentheses. Any instruction that would cause this limit to be exceeded instead acts as a `NOOP`, leaving all stacks in the states that they were in before the execution of the instruction.

The convention regarding the order of arguments for instructions that are more commonly rendered as infix operators is that the argument on the top of the stack is treated as the right-hand argument and the argument second-from the top is treated as the left-hand argument. This means that the linear representation of an expression containing one of these instructions looks like the normal infix expression, except that the instruction has been moved to the end. For example, we divide 3.14 by 1.23 using "`( 3.14 1.23 FLOAT./ )`". Similarly, 23 minus 2 is expressed as "`( 23 2 INTEGER.- )`".

While Push's stacks are generally treated as genuine stacks — that is, they are accessed only "last in, first out," with instructions taking their arguments from the tops of stacks and pushing their results onto the tops of stacks — a few instructions (like `YANK` and `SHOVE`) do allow direct access to "deep" stack elements by means of integer indices. To this extent the stacks can be used as general, random access memory structures. This is one of the features that ensures the Turing-completeness of Push (another being the arbitrary name/value bindings supported by the `NAME` data type and `SET/GET` methods).

Additional types can be added to a Push implementation with relative ease; types that have been added to date include vectors and unitary matrices (as described in Chapter 7), and one could add both additional standard types (for example, arrays and strings) and more exotic types (possibly URLs, images, etc.) for the sake of particular applications. A standardized interpreter configuration file format helps to ensure that different Push implementations can be configured to behave in the same ways on the same inputs.

More information on Push, including the current language specification document, pointers to implementations of Push interpreters (some in source code form), and related publications can be found online via the Push project home page.[2]

3. Push Examples

This section contains just a few simple examples, to give the reader a feel for the language and a few of its features that are convenient for genetic and evolutionary computation. More examples, including test suites, are available from the Push project home page.

[2] `http://hampshire.edu/lspector/push.html`

Function RANDOM-CODE (input: MAX-POINTS)

- Set ACTUAL-POINTS to a number between 1 and MAX-POINTS, chosen randomly with a uniform distribution.
- Return the result of RANDOM-CODE-WITH-SIZE called with input ACTUAL-POINTS.

Function RANDOM-CODE-WITH-SIZE (input: POINTS)

- If POINTS is 1 then choose a random element of the instruction set. If this is an ephemeral random constant then return a randomly chosen value of the appropriate type; otherwise return the chosen element.
- Otherwise set SIZES-THIS-LEVEL to the result of DECOMPOSE called with both inputs (POINTS − 1). Return a list containing the results, in random order, of RANDOM-CODE-WITH-SIZE called with all inputs in SIZES-THIS-LEVEL.

Function DECOMPOSE (inputs: NUMBER, MAX-PARTS)

- If NUMBER is 1 or MAX-PARTS is 1 then return a list containing NUMBER.
- Otherwise set THIS-PART to be a random number between 1 and (NUMBER − 1). Return a list containing THIS-PART and all of the items in the result of DECOMPOSE with inputs (NUMBER − THIS-PART) and (MAX-PARTS − 1)

Figure 6.2. The random code generation algorithm used both for the CODE.RAND instruction and for generating random programs for other purposes, for example in the initialization phase of PushGP.

First, some simple arithmetic:

```
( 5 1.23 INTEGER.+ ( 4 ) INTEGER.- 5.67 FLOAT.* )
```

.

Execution of this code leaves the relevant stacks in the following states:

```
FLOAT STACK: (6.9741)
CODE STACK: ( ( 5 1.23 INTEGER.+ ( 4 ) INTEGER.- 5.67
                FLOAT.* ) )
INTEGER STACK: (1)
```

A few points to note about this example:

- Operations on integers and on floating point numbers can be interleaved; all instructions take their arguments from the appropriate stacks and push their results onto the appropriate stacks.
- The call to `INTEGER.+` does nothing because there are not two integers on the `INTEGER` stack when it is executed.
- The call to `INTEGER.-` subtracts 4 (which is on top of the stack) from 5 (which is second on the stack), not the other way around.
- The parentheses in "`( 4 )`" have no effect on the results; parentheses serve mainly to group pieces of code for handling by code-manipulation instructions.

Here is a tiny program that adds an integer pre-loaded onto the stack to itself:

```
( INTEGER.DUP INTEGER.+ )
```

When run with 5 pre-loaded onto the `INTEGER` stack, for example, this leaves 10 on top of the stack. The following does the same thing in a slightly more complicated way, pushing code onto the `CODE` stack and then executing it:

```
( CODE.QUOTE ( INTEGER.DUP INTEGER.+ ) CODE.DO )
```

The "doubling subroutine" used in this example can be reused in a variety of ways. For example, one can use the `CODE.DUP` instruction to make multiple copies of the subroutine for multiple executions. In the following example the subroutine is duplicated twice, and then all three copies are executed sequentially via three calls to `CODE.DO`. When run with 5 pre-loaded onto the `INTEGER` stack this leaves 40 on top of the stack.

```
( CODE.QUOTE ( INTEGER.DUP INTEGER.+ )
  CODE.DUP CODE.DUP
  CODE.DO CODE.DO CODE.DO )
```

Another mechanism for code reuse involves named code variables. The following example is functionally equivalent to the one above, but the doubling subroutine is stored in the variable DOUBLE using CODE.SET and then retrieved multiple times using CODE.GET rather than being duplicated using CODE.DUP:

```
( CODE.QUOTE ( INTEGER.DUP INTEGER.+ ) DOUBLE CODE.SET
  DOUBLE CODE.GET CODE.DO
  DOUBLE CODE.GET CODE.DO
  DOUBLE CODE.GET CODE.DO )
```

Although the named subroutine technique is more verbose than the duplicated subroutine technique in this simple case, it may be convenient for storage of code during other tasks that also use the CODE stack.

The following more complicated example uses code duplication and also recursive calls (using CODE.DO in the subroutine) to compute the factorial of an integer pre-loaded onto the INTEGER stack. This example makes use of the fact that top-level calls to the interpreter push the executed code onto the CODE stack before execution:

```
( CODE.QUOTE ( INTEGER.POP 1 )
  CODE.QUOTE ( CODE.DUP INTEGER.DUP 1 INTEGER.- CODE.DO
               INTEGER.* )
  INTEGER.DUP 2 INTEGER.< CODE.IF )
```

This works by first pushing two pieces of code (for the base case and recursive case of the recursive factorial algorithm, respectively) onto the CODE stack; these are pushed on top of the code for the full program, which is pre-loaded onto the CODE stack by the top-level call to the interpreter. The subsequent code compares the provided integer with 2 and, depending on the result of this (which will be found on the BOOLEAN stack), executes one of the pushed pieces of code (and discards the other). In the base case this will produce an answer of 1, while in the recursive case it will recursively compute the factorial of one less than the provided number, and multiply that result by the provided number. When called with 5 pre-loaded on the INTEGER stack this leaves the relevant stacks in the following states:

```
CODE STACK: (( CODE.QUOTE ( INTEGER.POP 1 )
               CODE.QUOTE ( CODE.DUP INTEGER.DUP 1
                            INTEGER.- CODE.DO INTEGER.* )
               INTEGER.DUP 2 INTEGER.< CODE.IF ))
BOOLEAN STACK: ( )
INTEGER STACK: ( 120 )
```

A simpler implementation of a Push factorial function can be produced using the DO*COUNT iteration instruction, which is but one of several other instructions that recursively invoke the interpreter on code that is on the CODE stack. DO* is like DO except that it pops its code argument before, rather than after, the code argument is executed. DO*TIMES is like DO* except that it executes the popped code a number of times that is taken from the INTEGER stack. DO*COUNT is like DO*TIMES except that it also pushes an iteration counter (starting with 0) onto the INTEGER stack prior to each iteration. With DO*COUNT an iterative factorial function can be expressed as follows:

```
( CODE.QUOTE ( 1 INTEGER.+ INTEGER.* )
  1 INTEGER.SWAP CODE.DO*COUNT )
```

In all of the preceding examples, the pieces of code that were used as subroutines were simply copied (on the stack or via variables) and re-executed without alteration. But Push also includes a rich set of code-manipulation instructions that allow programs to modify code in arbitrary ways prior to execution. These include several instructions modeled on Lisp's list-manipulation functions (such as CODE.CAR, CODE.CDR, and CODE.CONS), along with special-purpose, higher-level instructions such as CODE.DISCREPANCY (which pushes a measure of the difference between the top two CODE stack items onto the INTEGER stack), CODE.RAND (which generates random code using the algorithm in Figure 6.2), and others. As an example of the sort of dynamic code construction that is possible, consider the following Push program for calculating 2^n for a positive value of n that is pre-loaded onto the INTEGER stack:

```
( CODE.QUOTE ( INTEGER.DUP INTEGER.+ ) DOUBLE CODE.SET
  CODE.QUOTE ( )
  CODE.QUOTE ( DOUBLE CODE.GET CODE.APPEND )
  CODE.DO*TIMES
  1 CODE.DO )
```

The first line of this program defines the same DOUBLE subroutine that was used in a previous example. Line 2 pushes an empty list onto the CODE stack. Line 3 pushes a piece of code that says "append the definition of DOUBLE to whatever is on top of the CODE stack." Line 4 pops the code from line 3 and then executes it n times (since n was pre-loaded onto the INTEGER stack). At the end of this step the top item of the CODE stack will be a program consisting of n repeated instances of "INTEGER.DUP INTEGER.+". For example, if 3 has been pre-loaded onto the INTEGER stack then the top item on the CODE stack after the execution of line 4 will be:

```
( INTEGER.DUP INTEGER.+
  INTEGER.DUP INTEGER.+
  INTEGER.DUP INTEGER.+ )
```

The 5th and final line of the program pushes 1 onto the INTEGER stack and then executes the program that was constructed by line 4. This leaves 2^n on top of the INTEGER stack. Although this is an unusual and somewhat verbose way of calculating 2^n, it nonetheless illustrates some of the ways in which Push programs can dynamically manipulate and execute code.

4. PushGP: Genetic Programming with Push

PushGP is a genetic programming system that evolves Push programs. It is a simple system in many respects, in part because it was initially designed merely as a demonstration of the use of Push in a genetic programming system. Most of its algorithms and features are the same as those used in the simplest traditional genetic programming systems, and some, owing to efficiencies made possible by Push's minimalist syntax, are even simpler. Nonetheless, the use of Push as the language in which evolving programs are expressed provides the following attractive features:

- Multiple data types without constraints on code generation or manipulation.
- Arbitrary modularity without constraints on code generation or manipulation.
- Evolved module architecture with no extra machinery.
- Support for explicit, arbitrary recursion.
- Support for code self-development and, via extensions such as Pushpop, the evolution of diversifying reproduction procedures (see Section 6.5).

PushGP is described in detail in documents available from the Push project's online web page.[3] Here we describe only its basic structure and note a few aspects of its performance.

A PushGP run begins with the generation of a population of random programs, using the algorithm shown in Figure 6.2. Each program in the population is then evaluated for fitness with respect to the target problem. If a solution to the target problem has been found then it is printed and the system halts. Otherwise a new generation of programs is produced through the application of genetic operators to programs in the current generation that are selected via fitness tournaments. These are then in turn evaluated for fitness, and the process continues until a solution is found or until a pre-established generation limit has been reached.

The genetic operators that are used in PushGP generally include exact reproduction and simple variants of the mutation and crossover operators that were described in Section 5.3, "liberalized" somewhat to suit the more permissive syntax of Push. Additional operators, such as an `ADD` operator that inserts a new subprogram within the parent and a `REMOVE` operator that deletes a subprogram from the parent, are also often used; these have no direct analogues in traditional genetic programming because such insertions and deletions would, if not performed carefully, produce programs that violate the argument-number requirements of Lisp-like representations. Alternative operators, including some designed to combat run-away code growth ("bloat"), have also been explored (Robinson, 2001; Crawford-Marks and Spector, 2002). In addition, an `IMMIGRATION` operator, which copies programs from disk files, is sometimes used to facilitate the use of multiple-deme evolution architectures across networks of workstations.

There is little remarkable about the overall PushGP algorithm itself; for the most part it is just a re-implementation of traditional genetic programming. But the fact that the underlying program representation is Push, which supports multi-type programs and complex control structures via code self-manipulation, means that this simple system can evolve multi-type, modular, recursive, and self-developing programs with no additional mechanisms.

The discerning reader will note that it is one thing to say that the Push representation "supports" various capabilities, and another thing entirely to demonstrate that programs with these capabilities actually do evolve in practice. It is yet another thing to demonstrate, for ex-

[3] `http://hampshire.edu/lspector/push.html`

ample, that the problem-solving advantages of modularization schemes such as ADFs are also obtained from the "emergent modularization" process that can occur during a PushGP run. Such demonstrations are important, but beyond the scope of this book. The fact that many of the results in Chapter 8 were produced with PushGP provides some anecdotal evidence of PushGP's efficacy, but it does not constitute a systematic assessment. Such assessments have been published elsewhere demonstrating, for example, "scale up" properties better than those of genetic programming with ADFs on parity problems (Spector and Robinson, 2002a), and modularity-based robustness properties similar to those provided by genetic programming with ADFs on a problem in a dynamic environment (Spector and Robinson, 2002b).

PushGP often finds unexpected ways to leverage the code manipulation and multi-type facilities of Push to produce unusual (and "un-human-like") solutions. For example, while PushGP routinely produces recursive code, the code that is produced rarely follows the neat outlines recommended in programming textbooks, and considerable effort is sometimes required to understand the ways in which evolved Push programs manipulate and execute code to achieve their results. In one example, presented in (Spector and Robinson, 2002a), a program evolved to solve the ODD problem, of determining whether its input is or is not an odd number, did so by using the provided number as an index into the program itself, and by evaluating a property of the code found at that index. This was a clever (and 100% correct) solution, but not one that a human programmer would be likely to devise.

PushGP is "self-adaptive" insofar as the number and architecture of modules to be used in a solution, along with the selection of data types to be employed, are determined dynamically and automatically as part of the evolutionary process. To some extent the highly redundant nature of Push syntax — that is, the facts that parentheses can often be added or deleted and that instruction sequences can often be reordered, all without changing the function of a program — allows for other forms of (representational) self-adaptation during a run. Many aspects of the system, however, must be specified or adjusted manually. For example, PushGP uses hand-designed mutation and crossover algorithms and hand-specified rates of application for each of the genetic operators. Beyond PushGP, the code-manipulation features of Push can support more radically self-adaptive forms of genetic and evolutionary computation, in which more aspects of the system are under evolutionary control. Some of these self-adaptive extensions are described in the following section.

5. Autoconstructive Evolution

Push allows one to integrate, in a syntactically uniform way, the manipulation of code with the manipulation of problem-oriented data types (integers, floating point numbers, matrices, etc.). As shown in the previous section, this capability can be used to support the evolution of programs that use modules and novel control structures. But it can also be used, more ambitiously, to bring more of the evolutionary process under evolutionary control. It be used in this way because a genetic and evolutionary computation system is *itself* made of code, some of which may also be permitted to evolve. This section briefly describes some of the ways in which these ideas can be applied with Push. Note, however, that the relatively standard PushGP system described in the previous section has thus far proven more useful in problem-solving contexts (including automatic quantum computer programming, as described in Chapter 8) than have such "meta-evolutionary" systems. But it was to support such systems that Push was originally designed, and these systems may, by virtue of their self-adaptive capabilities, produce even more powerful problem-solving technologies in the future.

Several previous genetic and evolutionary computation systems have incorporated some form of self-adaptation such as the genetic encoding of mutation rates. In the genetic programming literature more specifically, several "Meta-GP" systems have been described in which the rates and also the algorithms for mutation are genetically encoded and therefore subject to evolution (Schmidhuber, 1987; Edmonds, 2001). In place of traditional genetic operators these systems use co-evolving populations of program-manipulation programs to produce the offspring of the individuals in the primary (problem-solving) population.

An "autoconstructive evolution" system is a genetic and evolutionary computation system in which the evolving problem-solving programs are *themselves* responsible for the production (and diversification) of their own offspring, just as biological organisms are responsible both for "making a living" in their environments and for producing their own offspring (Spector, 2001; Spector and Robinson, 2002a). Since the means by which programs create their offspring are embedded within the evolving programs themselves, and are therefore subject to variation and natural selection, significant aspects of the evolutionary process thereby come under evolutionary control. As an autoconstructive evolution system runs, the evolutionary process itself, insofar as it is implemented by the reproductive behaviors of the evolving programs, is constructed by evolution.

Autoconstructive evolution can be accomplished with Push in a variety of ways. The first Push-based autoconstructive evolution system, Pushpop, was derived via minimal changes to PushGP. An additional code data type, called CHILD, was added that supported all of the standard code-manipulation instructions except those that cause recursive execution (like DO, DO*, and IF). At the end of each fitness test any code left on the top of the CHILD stack became a potential child, stored with the parent (which could then be considered to be "pregnant") until the selection phase. In the selection phase, after all programs in the population had been evaluated for fitness (and had produced their potential children), fitness tournaments between the parents determined from which parents children would be taken for inclusion in the next generation.

To avoid the evolutionary stagnation that would result from programs that produced only exact clones of themselves, a "no cloning" rule was imposed; children were not added to the subsequent generation if they were duplicates either of their parent or of other children that had already been added.[4] Aside from the prohibition against clones, programs could produce their children in any manner that was expressible in Push code, including standard mutation and crossover procedures as special cases. To support sexual recombination procedures (such as crossover), special instructions were provided to access other individuals in the population (selected by distance, parent's fitness, or program contents) and to push their programs onto the current CODE stack. Note that code could be "borrowed" from more than one mate, enabling complex forms of multi-way recombination. Because this "borrowed code" could be used not only for the production of children, but could also be executed by the parent and thereby contribute to the problem-solving behavior of the parent, the programs in a Pushpop population could become tightly interdependent.

Pushpop has been demonstrated to solve simple symbolic regression problems, but its primary utility to date has been in the study of self-adaptive evolutionary processes themselves. For example, it has been used to explore relations between diversification and adaptation, showing that adaptive populations of Pushpop programs are reliably more diverse than required by the "no cloning" rule (Spector, 2002).

It is possible to take additional steps in this self-adaptive direction, for example to allow reproductive timing to be controlled by individual programs (and hence by evolution), rather than by the hand-designed,

[4]There is no relation between Pushpop's "no cloning" rule and the "no cloning theorem" of quantum information theory.

generation-based schemes of both PushGP and Pushpop. A system called SWARMEVOLVE 2.0 does this, although in a rather different architectural context. SWARMEVOLVE 2.0 is an autoconstructive evolution system in which flying agents evolve in a 3D virtual world that is implemented in the BREVE simulation environment (Klein, 2002).[5] These agents decide on their own when, where, and how to produce children (using a `SPAWN` instruction), and many more features of the evolutionary system are thereby under evolutionary control. This system has been used primarily to study evolutionary dynamics; for example, it has served as a framework for exploring the evolution of collective behaviors (Spector and Klein, 2002; Spector et al., 2003a; Spector et al., in press).

The utility of autoconstructive evolution systems for automatic quantum computer programming is currently unknown. One may speculate, however, that such systems will eventually out-perform traditional genetic programming systems by adapting their reproductive mechanisms and their representations to their problem environments.

[5]BREVE is available from `http://www.spiderland.org/breve`.

generation-based schemes of both Bratko and Pachpec. A [illegible] called SWARMEVOLVE 2 [illegible] [illegible]. SWARMEVOLVE 2.0 [illegible] autoconstructive evolution [illegible] virtual world [illegible] BREVE simulation environment (Klein, 2002). These [illegible] and how to produce and/or [illegible] features of the evolution [illegible]. The systems [illegible] evolutions [illegible] [illegible] [illegible] [illegible] [illegible]

[1] [illegible]

Chapter 7

EVOLUTION OF QUANTUM PROGRAMS

This chapter presents specific strategies for the evolution of quantum programs using the technologies presented earlier in this book. The application of these strategies to particular problems is documented in Chapter 8. Related strategies have also been developed and applied by other researchers (for example Williams and Gray, 1999; Surkan and Khuskivadze, 2001; Leier and Banzhaf, 2003a; Leier and Banzhaf, 2003b; Perkowski et al., 2003; Massey et al., 2004); while some of these efforts are cited in the following discussion, the focus here is on the strategies that have been developed by the author.

1. Program Representations

A genetic programming system can be thought of as a search procedure that searches the "space" of computer programs to find a program that meets some particular, usually behavioral, criterion. The search space is normally infinite, meaning that no finite search procedure can explore it completely. If one considers programs with real-valued parameters (like many quantum programs) then the search space is *uncountably* infinite, even for programs of finite size. Even when one substitutes limited-precision floating point numbers for true real numbers, the number of programs of any nontrivial finite size is astronomically large. It is therefore important for the designers of genetic programming systems to carefully consider the subset of the space that will be examined by their methods. The membership of this subset is influenced by many factors ranging from program representation to the choice of genetic operators.

In traditional genetic programming the "chromosomes" upon which the Darwinian processes of variation and natural selection act are the programs themselves. That is, one determines the behavior of an individual with a particular chromosome by executing the chromosome, which is in fact the whole of the individual. By contrast, as described in Chapter 6, in developmental approaches the execution of the chromosomal program produces *another* program which is then executed to produce the problem-solving behavior. Both the direct, chromosomal encoding and the indirect, developmental encoding of quantum programs can be used for quantum program evolution.

In either case one must first determine the forms that the evolved and/or developed quantum program will take. For some problems one might have pre-established constraints that can be used to design effective program representations. For example, if one knew that the program that would solve a particular problem takes the form of a single-qubit gate then one could represent the program using the four real-valued parameters of a $U2$ gate (see Chapter 2). One could then attempt to evolve the problem-solving single-gate program by using chromosomes consisting of the four real-valued parameters, and genetic operators that operate on strings of four numbers. Alternatively, one could use chromosomes of some more elaborate form such as programs which, when executed, set the values of the four parameters.

For most interesting problems we seek programs that operate on many qubits. Although it would be possible in principle to represent these larger quantum programs using only numerical parameters, extending the idea described in the previous paragraph, the number of parameters grows rapidly as the size of the program increases. Worse, the *meanings* of the large number of parameters become difficult for humans to discern, so that the results of an evolutionary process producing sets of such parameters would be difficult to analyze or extend. For these reasons it often makes more sense to represent quantum programs as sequences of well-understood quantum gates that operate on small numbers of qubits. The QGAME representation for quantum programs, described in Chapter 3, was designed for this purpose.

Once the representational scheme for the quantum programs has been determined — and for the remainder of this chapter it will be assumed that quantum programs take the form of QGAME programs — one still faces choices with respect to chromosome representation. One possibility is to use the QGAME programs themselves as chromosomes. This is essentially the technique that was described as "stackless linear genome genetic programming" with "encapsulated gates" in (Spector et al., 1999a; Spector et al., 1999b), and it is similar to the techniques that have been used by several other researchers (Williams and Gray, 1999; Leier and Banzhaf, 2003a; Massey et al., 2004). Because QGAME programs are syntactically unconstrained — that is, any re-ordering of the instruction expressions in a QGAME program yields another syntactically well-formed QGAME program — simple, "blind" genetic operators can be used in conjunction with these chromosomes. Several of the results presented in Chapter 8 were first obtained using this technique.

There are several reasons, however, that one might wish to use a more expressive, and in fact developmental, chromosome representation. For example, one may expect that the quantum programs that are solutions to some problems will include multiple instances of the same parameter value (for example, as a parameter to `U-THETA`[1]), and it would be desirable to allow a single instance of such a parameter in a chromosome to "translate" into several instances of the parameter in the problem-solving quantum program. In other cases the problem itself may have parameters (for example, a number of times that an oracle gate may be called), and it would be desirable to allow this parameter to directly influence the construction of the problem-solving quantum program. Many important problems have this property because they involve the discovery of *scalable* quantum programs that can solve problems of various sizes. One strategy for solving these problems is to provide a size parameter that influences the form of the resulting quantum program.[2] Finally, in some cases it may be useful to allow components of evolved quantum programs to be produced via computational manipulations of other components of the same programs; for example, it may be useful to allow the angle of a `U-THETA` gate to be obtained from the difference between the angles two other `U-THETA` gates.

A developmental genetic programming approach, as described in Chapter 6, can provide all of these capabilities. In this approach the chromosomal programs are expressed in some classical (non-quantum) programming language and may take various forms including Lisp-style program trees or stack-based instruction sequences. These programs, when executed, *construct* the problem-solving quantum program.[3] The chromosomal programs may include both classical instructions, which manipulate

[1]See Chapter 2 for a description of QGAME instructions and their associated matrices.
[2]An example of the evolution of a scalable quantum program is presented in Chapter 8.
[3]A similar approach is explored in (Massey et al., 2004).

standard data types such as numbers and Booleans, and instructions that add components to a developing QGAME program. Before the execution of the chromosomal program one initializes an "embryo" that consists of a minimal QGAME program; this embryo may be a completely empty program or it may contain, for example, instructions that initialize the states of certain qubits or conduct final measurements. The execution of a quantum-component-adding instruction in the chromosomal program augments the embryo with a specific QGAME instruction expression. These instruction expressions are typically added to the end of the developing QGAME program, before any pre-specified final measurement gates.

The use of Push and PushGP (as described in Chapter 6) for such a developmental approach provides several advantages. For example, the ease with which multiple data types can be integrated into Push allows one to add a `QGATE` type that supports complex evolved strategies for quantum program development. Data of this type, as implemented in the version of PushGP that was used to produce several of the results in Chapter 8, consists of fully expanded unitary matrices (for quantum systems of a pre-specified size) along with "history" specifications that show how the matrices were constructed from primitive gate matrices. Most of the `QGATE` instructions expand a particular matrix to the requisite size, taking the arguments that they need to do so from the appropriate stacks (and taking qubit indices modulo the number of qubits), and push the resulting `QGATE` structure onto the `QGATE` stack.[4] For example, the `QGATE.HADAMARD` instruction, when executed in the context of a 2-qubit quantum computer and with 0 on the `INTEGER` stack, pushes a structure with the following unitary matrix (in which all numbers have been rounded to 4 decimal places for readibilty) onto the `QGATE` stack:

```
((0.7071 0.7071 0.0000 0.0000)
 (0.7071 -0.7071 0.0000 0.0000)
 (0.0000 0.0000 0.7071 0.7071)
 (0.0000 0.0000 0.7071 -0.7071))
```

The history attached to this matrix is simply `(HARAMARD 0)`. The subsequent execution of a `QGATE.GATE` instruction would add a QGAME instruction (in this case one that uses QGAME's `MATRIX-GATE` construction with the matrix specified above) to the developing QGAME program.

[4] `QGATE` instructions corresponding to non-unitary QGAME program elements such as `MEASURE`, `END`, and `HALT`, along with instructions that produce oracle calls, bypass the `QGATE` stack and directly augment the developing QGAME program.

Before the execution of QGATE.GATE, however, it is possible for the Push program to store and to transform the unitary matrix. For example, consider the following Push program fragment:

```
0 QGATE.HADAMARD 1 0.23 QGATE.U-THETA
QGATE.COMPOSE 0 1 QGATE.CNOT QGATE.COMPOSE
QGATE.TRANSPOSE
```

This code first pushes the same matrix that was used in the previous example onto the QGATE stack. It then pushes a U-THETA matrix (applied to qubit 1, with angle 0.23) on top of the HADAMARD matrix. The subsequent QGATE.COMPOSE instruction pops both matrices from the stack and replaces them with their composition (and a history that reflects the origins of the composition).[5] The next three items push a CNOT matrix (with qubit 0 as the control and qubit 1 as the target), and the subsequent QGATE.COMPOSE instruction composes the previously constructed matrix and the CNOT. The call to QGATE.TRANSPOSE transposes the result of the final composition, producing the following matrix (rounded):

```
((0.0688 0.0688 -0.0161 -0.0161)
 (0.0161 -0.0161 0.0688 -0.0688)
 (0.0161 0.0161 0.0688 0.0688)
 (0.0688 -0.0688 -0.0161 0.0161))
```

This matrix could then be added to a developing QGAME program by means of a call to QGATE.GATE, although it (or its components) could also be duplicated via QGATE.DUP or stored in a named variable. It could thereby be used multiple times, with or without further manipulation, in the fully developed quantum program. The code-manipulation features of Push and PushGP that allow for the emergence of modules and other control structures during evolution can leverage this manipulation of the QGATE stack to ease the evolution of quantum programs with complex, modular structures.

[5]Because computer representations of quantum gates often include small round-off errors it is important to check for unitarity when composing large numbers of gates; otherwise the accumulated round-off errors may significantly violate the unitarity constraint and the transformations specified by the composed gate may correspond to physically impossible operations. It is also sometimes useful to limit the amount of composition for other reasons, for example to facilitate human analysis; to this end the gate composition procedures supported by QGAME, and utilized by Push instructions like QGATE.COMPOSE, will refuse to compose gates if their combined histories exceed a pre-specified nesting depth.

2. Fitness

The genetic programming process requires all individuals in the evolving population to be assessed for fitness each generation. How might this fitness assessment be performed for the quantum programs that are produced by the developmental processes described in the previous section?

In some cases one might be able to use special-purpose fitness measures that avoid the need for actual quantum computer simulation. For example, for some problems that use no non-unitary program elements (such as measurements), it may be possible to build the composite unitary matrix represented by an entire QGAME program and to directly assess features of this composite matrix that are relevant to the fitness of the program for the problem in question. This was the approach taken by Colin Williams and Alexander Gray in their work using genetic programming to find decompositions of pre-specified unitary matrices; they built the composite matrices and compared them, element by element, with the target matrix (Williams and Gray, 1999). In most cases, however, this approach is impractical, either because the problem calls for non-unitary elements, or because the relevant features of the composite matrix are not easy to assess, or because the use of oracle gates would mandate the construction of a large number of composite matrices. For these reasons it is often simpler to simulate the execution of the program on a quantum computer, and to compute the program's fitness from the simulation output.

As discussed in Chapter 2, a quantum computer running a particular quantum program may produce different outputs, each with a particular probability, from successive but otherwise identical runs. If we were to use a *real* quantum computer to assess fitness then we would only get one output from each run, and we would have to run the program many times to determine the probabilities for each output. With quantum computer simulators such as QGAME, however, we receive a list of all possible outputs and their associated probabilities from a single run (albeit from a run that requires exponential computational resources). All of this information about possible outputs and their probabilities can be used in a fitness function for genetic programming.

The most straightforward way to use this information in a fitness function is to use the probability of error directly as a fitness value, with a fitness of zero indicating a perfect solution. This is similar to the use of an error value as a fitness value in traditional symbolic regression problems (as described in Chapter 5), although the *probability* of error is here being used in place of an actual numerical error. For a problem that has multiple fitness cases we might calculate fitness as the sum or average of the individual probabilities of error, or, depending on the requirements of the problem, as the maximum probability of error for any particular fitness case.

Probability of error is a useful measure of quantum program quality, but it is often insufficient, by itself, as a fitness measure for quantum program evolution. This is because it is often extremely easy to produce a program with a 50% probability of error that has no resemblance to a true solution. This can often be done simply by rotating the output qubits to equal superpositions of 0 and 1 (as might be done, for example, with a `HADAMARD` gate). Programs that are "better" than this, in the sense that they contain more components of true solutions, may have a lower probability of error on some fitness cases but a higher probability of error on others, and their average probabilities of error may also be higher than 50%. As a consequence, programs that achieve a 50% probability of error often form a troublesome local minimum in the search space produced by a fitness function that considers only probability of error.

For this reason it is often useful to consider, in addition to the total, average, or maximum probability of error, a measure of the number of fitness cases for which the program is more likely than not to produce the wrong answer. This measure will be called the number of "misses" here.[6] To ensure that one counts as misses even those cases that only dip below 50% probability of error because of round-off errors, one should generally make the threshold for a miss somewhat lower; in the examples presented in Chapter 8 the threshold used was always 48%. As described in Chapter 2, QGAME's `TEST-QUANTUM-PROGRAM` function returns the number of misses along with error probability and oracle statistics.

One might combine probability of error with misses to produce an evolution-guiding fitness function in a variety of ways. In many cases a "lexicographic" combination, in which the probability of error serves only to distinguish among programs with an identical number of misses, can be effective. For example, suppose we are conducting a run with four fitness cases, and that we use $f = 10m + e$ as the fitness function, where m is the number of misses and e is the maximum probability of error on any one case. The first term of the sum will be one of 0, 10, 20, 30, or 40, while the second term will be a real number less than or equal to 1. As previously, lower fitness values are considered better. The result will be that a program that achieves a lower number of misses will always be considered better than a program that achieves a higher number of misses. When comparing two programs that achieve the same number

[6]This usage should not be confused with John Koza's use of the term "hits," which is only loosely (and inversely) analogous (Koza, 1992).

of misses the one with the lower maximum probability of error will be considered better.

Many other combination methods are possible, including some studied for other kinds of genetic programming problems and for "multi-objective optimization problems" more generally (see, for example, Ekart and Nemeth, 2001). In addition, certain problems might call for unique manipulations of error probabilities and counts of misses. For example, for some problems the coarse, discrete nature of the misses count may produce problematic "plateaus" in the fitness landscape; that is, it might be necessary to temporarily explore programs with higher numbers of misses (but perhaps lower probability of error) in order to make progress, but this exploration may be made impossible by the "all or nothing" quality of each "miss," combined with lexicographic combination of misses and probability of error. In such cases one might try "smoothing" the misses count (for example, with a sigmoid function centered on 48%) or transforming misses and/or error probabilities in some other way. There are no definitive guidelines on how best to do this for any particular problem, but analysis of failed runs can sometimes lead to interventions that eventually allow the system to discover a solution.

Additional fitness components may be included for particular problems. For some problems it may make sense to allow quantum programs to include multiple calls to an oracle gate, and it may be desirable to minimize the number of such calls by the end of the run; for these problems it might be useful to combine the number of oracle calls with other values in the fitness function. One might also wish to minimize the *total* number of gates; this measure could be included in the fitness function as a sort of "parsimony" component (Koza, 1992). It might also be useful in some cases to include measures of quantum mechanical properties such as entanglement in the fitness function, either because they are directly relevant to the problem being solved or because they are expected to have some particular values in solutions.

3. Operators and Refinements

As described in the previous sections, the application of genetic programing to the task of automatic quantum computer programming calls for certain design decisions to be made with respect to program representations and fitness functions. Such decisions are among the preparatory steps that must be taken in any application of genetic programming. This section briefly describes a few additional refinements which have been employed by the author in the evolution of certain quantum programs, and which may also be useful in other situations.

Some of these refinements concern the genetic operators used to mutate and recombine the chromosomal programs. While the standard genetic operators provided with PushGP and with other genetic programming systems will suffice for many problems, additional operators that are specialized for quantum program evolution may also be useful. For example, because small changes to angles used in rotation gates often result in significant changes in the behavior of the programs within which they are embedded, it may be useful to include a "number mutation" operator that adds small random numbers (generated, perhaps, via Gaussian noise) to the floating-point literals in a Push program. Such operators have been used previously in other forms of genetic and evolutionary computation (see, for example, Fogel and Atmar, 1990), but they are not typically used in genetic programming systems.

The compositional properties of unitary gates suggest additional genetic operator refinements that are uniquely applicable to quantum program evolution. In approaches that use QGAME programs or similar representations as chromosomes one can use a "gate compression" operator which composes a sequence of unitary gates into a single gate, as described in conjunction with the `QGATE.COMPOSE` Push instruction above. This compresses a program, or a segment of a program, and produces in its place a complex single gate, functionally equivalent to the entire segment, which may later be further compressed and/or passed to other programs via recombination. This scheme is similar in some respects to techniques developed for building libraries of subroutines in traditional genetic programming (Koza, 1990; Koza, 1992; Angeline and Pollack, 1992), although the compression of multiple gates into a single unitary matrix (which can be simulated with no greater cost than any other single matrix) has no direct analogue in most classical systems. As in all applications of unitary matrix composition one should be careful to avoid the accumulation of round-off errors that would produce non-unitary (and hence physically impossible) results.

In developmental approaches, gate compression cannot be performed directly as a genetic operator, because the chromosome does not itself contain sequences of gates, but rather code that produces sequences of gates when executed. However, a related process of "matrix literalization" may sometimes be useful.

Matrix literalization takes place after fitness testing and is applied to some small number of high-performing individuals. The chromosomal programs of these individuals are re-executed to produce QGAME programs, and then gate compression is performed on the QGAME programs to produce matrices that compute significant segments of the more-successful quantum programs. These compressed matrices are then made into QGATE literals that may later be included, via mutations, in other programs in the PushGP population. Over the course of evolution the population may come to include matrix literals, which may become hierarchically composed, that represent useful complex operations distilled into compact subroutine-like modules. Examples of matrices produced by this process are presented in Chapter 8.

Several genetic programming initialization steps may also be refined for application to problems in quantum computing, to increase the quantity of useful program structures in the initial population. For example, in most quantum oracle problems one knows in advance that a quantum program containing *no* oracle calls will be useless. One might therefore arrange for all randomly generated programs to include instructions that generate oracle gates; this might be accomplished either by discarding and regenerating any programs without the relevant calls, or by inserting such calls into all programs after random generation. Similarly, for such problems one can avoid wasting time during fitness evaluation by skipping the evaluation of any "clearly useless" programs; this can be accomplished at the level of the chromosomal program (for example by refusing to evaluate any Push program that contains no oracle-generating instructions), at the level of the quantum program (for example by refusing to evaluate any QGAME program that does not contain oracle gates), or at both levels. In all such cases the useless program should then be assigned a fitness penalty that ensures that it will lose all fitness tournaments to any programs that are not obviously useless.

Other initialization refinements concern the selection of constants and instructions that are available for inclusion in random programs. For example, because many known quantum algorithms involve the use of U-THETA rotations using values of θ that are ratios of π, it may in some cases be beneficial to use a specialized random floating point number generator that produces all or mostly numbers of this form. The selection of instructions can influence the performance of the system in many and complex ways, and although it is difficult to predict these influences in general, it may sometimes be possible to translate the requirements of a particular problem into a selection of instructions that improves the chances of finding a solution.

A final set of initialization refinements concerns the "embryo" from which evolved quantum programs develop. Various kinds of information that the developer might have about the desired quantum program can be included in this embryo, eliminating the need for evolution to rediscover this information. For example, in some cases one might know that solutions will involve the generation of an equal superposition of 0 and 1 for some set of qubits prior to the execution of the main part of the program; in such cases one might add a sequence of `HADAMARD` gates to the embryo, eliminating the need for these gates to be generated by the execution of the chromosomal program.

A few additional refinements concern the developmental process by which the problem-solving QGAME programs are produced. For evolving scalable quantum programs it is natural to include size-related instructions in the instruction set; for example, an example shown in the next chapter was evolved using a `NUMQUBITS` instruction that pushes the number of qubits in the quantum system onto the `INTEGER` stack. This instruction is used in the evolved Push program to control the developmental process, producing a larger QGAME program for larger numbers of qubits. Depending on the problem, one might want to provide other mechanisms to facilitate control of development; for example, it may sometimes be useful to include an iteration structure that iterates once for each qubit (or for each input qubit or for each output qubit), eliminating the need to combine independent calls to `NUMQUBITS` and the generic iteration structures.

Finally, in some cases it is useful to refine the developmental process by prohibiting certain additions to the developing quantum program. For example, in some of the problems presented in the next chapter the task is to determine whether information can be communicated between two sets of qubits by means of a single, particular gate that connects the two sets. The only quantum programs in which we are interested for this problem are those that connect the two sets of qubits with one and only one call to the particular gate under investigation. To produce such programs, and only such programs, we refine the developmental process to ignore all calls to add gates that would violate the communication restrictions. Other types of problems may involve different restrictions that can be handled in a similar way.

Chapter 8

EVOLVED QUANTUM PROGRAMS

This chapter presents examples of the automatic production of quantum computer programs via genetic programming. These examples demonstrate how the techniques described in previous chapters can be applied to specific problems. They also provide evidence for the claim that scientifically significant results can be produced via automatic quantum computer programming.

The examples that are presented here are solutions to two types of problems. We call problems of the first type "Boolean oracle analysis" problems because they require us to determine some property of a provided Boolean quantum gate. This gate is often called an "oracle" or a "black box" because we are given little *a priori* information about the gate's construction or behavior. All of these oracles are "Boolean" in the sense that they act by inverting a particular single output qubit when provided with specified combinations of inputs. We are allowed to use the oracle gate, but we are not told in advance which combinations of inputs will produce the inversion — that is what a solution to the problem will tell us. Sometimes we may be "promised" that the oracle is one of some subset of the possible Boolean oracles of the given size; in these cases the problem is to determine *which* member of the subset we have been given.

An example of a Boolean oracle analysis problem is Grover's database search problem, which was discussed earlier in Chapter 2. In Grover's problem the oracle represents a database containing a single "marked" item. We are promised that the oracle inverts its output for a single combination of inputs, which may be considered the address of the marked item. Our task is to determine which of the possible inputs it is for which the inversion is performed.

Other examples presented below — the Deutsch-Jozsa (XOR) problem, the Majority-ON problem, and the OR and AND/OR problems — are similar except that the "promises" that we are given about the oracles and the features of the oracles that we are asked to determine vary from problem to problem. For the Majority-ON problem we attempt not just to solve a single instance of the problem but rather to produce a scaling program that can solve instances of this problem of any size.

Several of these Boolean oracle analysis problems have practical significance because their solutions directly enable us to solve difficult real-world problems more rapidly than is possible on classical computers; for example, Grover's algorithm can be used to provide a quadratic speedup for a host of problems that involve search through unstructured databases.

The second type of problem considered here concerns the classical communication capacity of certain specific quantum gates. The problems of this type that are presented derive from recent research on the tradeoffs between classical communication and entanglement-generating powers of certain unitary transformations (Spector and Bernstein, 2003; Bennett et al., 2004). In these problems the task is to transfer information from one set of qubits to another, without any direct connection between the two sets of qubits aside from a single instance of the gate under investigation. These problems are important not because they have any direct practical application — the gates under consideration do not generally correspond to any real-world communication channels — but rather because their solutions contribute to the development of the fundamental theory of quantum communication and computation.

Sections 8.1 through 8.5 describe specific problems, specific genetic programming techniques that have been used to solve them, and interesting features of evolved solutions. Particular emphasis is given to the author's techniques described in Chapters 6 and 7 as they have been applied in specific cases. Section 8.6 discusses the general significance of the results presented in Sections 8.1 through 8.5, both with respect to the theory of quantum computation and with respect to techniques for automatic quantum computer programming.

1. The 1-bit Deutsch-Jozsa (XOR) Problem

In the Deutsch-Jozsa problem (Deutsch and Jozsa, 1992) we are given an oracle with some number of input qubits and one output qubit. We are told that the oracle's function is to invert its output qubit in certain situations (that is, with certain Boolean inputs), and we are promised that the oracle is either *uniform*, meaning that it either *always* or *never* inverts its output qubit, or *balanced*, meaning that it will invert and

Table 8.1. Push interpreter parameters for the example run of PushGP on the Deutsch-Jozsa (XOR) problem. Documentation of Push parameters and instructions is available from `http://hampshire.edu/lspector/push.html`.

MAX-RANDOM-FLOAT	1.0
MIN-RANDOM-FLOAT	-1.0
MAX-RANDOM-INTEGER	10
MIN-RANDOM-INTEGER	-10
EVALPUSH-LIMIT	150
MAX-POINTS-IN-RANDOM-EXPRESSIONS	50
MAX-POINTS-IN-PROGRAM	100
MAX-ORACLE-CALLS	1
Types	QGATE, FLOAT, CODE, BOOLEAN, INTEGER
Instructions	(see Table 8.3)

not invert *equal numbers of times* if called on all possible (Boolean) inputs. The task is to determine whether a given oracle is uniform or balanced. Classically one would have to query the oracle several times (up to one more than half the number of possible inputs) to be certain of the answer, but quantum computers can do better. Although this problem is not clearly related to any problems of practical significance, it is of historical significance because it was one of the first problems to be shown to be solvable with a better-than-classical quantum algorithm.

The use of genetic programming to re-discover the quantum program that solves the 2-bit version of this problem (which uses an oracle with 4 possible inputs) is documented in (Spector et al., 1998) and (Spector et al., 1999b).[1] Here we document the use of genetic programming to re-discover the quantum program that solves the simpler 1-bit version of this problem. In this version of the problem the oracle has only 1 input qubit and hence two possible inputs (0 and 1). The oracle is uniform, as in the general case, if it either always or never inverts its output qubit. It is balanced in all other cases, in which it inverts its output qubit for one but not the other of its 2 possible inputs. We are therefore asked to determine the truth of the logical formula $I_0 \oplus I_1$, where I_0 means "inverts with input 0," I_1 means "inverts with input 1," and $\oplus$ is the exclusive OR (XOR) function. The classical version of this problem clearly requires two oracle queries; after a query with one input it will not be known whether the result of a query with the other input will match (meaning that the oracle is uniform) or not (meaning that

[1] In these references the Deutsch-Jozsa problem is referred to as Deutsch's "early promise" problem.

the oracle is balanced). By contrast a quantum program can solve this problem with a single query.

This problem was easily solved using PushGP with the parameters shown in Tables 8.1 and 8.2 and the instruction set shown in Table 8.3, running under the OPENMCL open source Common Lisp system[2] on a 1.33 GHz Apple Macintosh laptop computer with a PowerPC G4 chip. The complete source code for this run, along with the output log, is available online.[3]

The fitness of a Push program was assessed by running it once to produce a QGAME program (which began with the empty "embryo" corresponding to the gate array shown in Figure 8.1), and by testing the QGAME program with the `TEST-QUANTUM-PROGRAM` function described in Chapter 3. The maximum permitted number of oracle calls per case (and therefore the first argument in all calls to `LIMITED-ORACLE`) was 1, so that only the first oracle call in any developed QGAME program would have any effect. The inputs provided to `TEST-QUANTUM-PROGRAM` were:

- `PROGRAM`: The developmental result of executing the chromosomal Push program.
- `NUM-QUBITS`: 2
- `CASES`: (((0 0) 0) ((0 1) 1) ((1 0) 1) ((1 1) 0))
- `FINAL-MEASUREMENT-QUBITS`: (1)
- `THRESHOLD`: 0.48

Fitness was computed as the sum of the number of misses (the first return value from `TEST-QUANTUM-PROGRAM`) and the maximum probability of error on any single case (the second return value).

The fitness of the best program in the first, random generation ("generation 0") was 3.0. Fitness improved rapidly thereafter, including a steep drop at generation 9 when the number of misses of the best program dropped from 2 to 0. At generation 18 a perfect solution was found, with a fitness value of 0 aside from a miniscule round-off error of 4.4×10^{-16}. A plot of the fitness of the best individual per generation is shown in Figure 8.2.

---

[2] `http://openmcl.clozure.com/`

[3] `http://hampshire.edu/lspector/aqcp/evolved-xor/`

Table 8.2. PushGP genetic programming system parameters for the example run of PushGP on the Deutsch-Jozsa (XOR) problem.

MAX-NEW-POINTS-IN-MUTANTS	20
POPULATION-SIZE	10,000
TOURNAMENT-SIZE	7
MUTATION-PROBABILITY	0.45
CROSSOVER-PROBABILITY	0.45
MUTATION-OPERATORS	FAIR, PERTURB, ADD, REMOVE
CROSSOVER-OPERATORS	FAIR
FITNESS-FUNCTION	misses + max probability of error

Table 8.3. Instructions used in the example run of PushGP on the 1-bit Deutsch-Jozsa (XOR) problem.

INTEGER	INTEGER.FROMBOOLEAN, INTEGER.FROMFLOAT, INTEGER.>, INTEGER.<, INTEGER.%, INTEGER./, INTEGER.*, INTEGER.-, INTEGER.+, INTEGER.STACKDEPTH, INTEGER.SHOVE, INTEGER.YANKDUP, INTEGER.YANK, INTEGER.=, INTEGER.SWAP, INTEGER.POP, INTEGER.DUP
BOOLEAN	BOOLEAN.FROMFLOAT, BOOLEAN.FROMINTEGER, BOOLEAN.NOT, BOOLEAN.OR, BOOLEAN.AND, BOOLEAN.STACKDEPTH, BOOLEAN.SHOVE, BOOLEAN.YANKDUP, BOOLEAN.YANK, BOOLEAN.=, BOOLEAN.SWAP, BOOLEAN.POP, BOOLEAN.DUP
CODE	CODE.DISCREPANCY, CODE.DO, CODE.NTHCDR, CODE.NTH, CODE.APPEND, CODE.LIST, CODE.NOOP, CODE.IF, CODE.DO*, CODE.CONS, CODE.CDR, CODE.CAR, CODE.NULL, CODE.ATOM, CODE.QUOTE, CODE.STACKDEPTH, CODE.SHOVE, CODE.YANKDUP, CODE.YANK, CODE.=, CODE.SWAP, CODE.POP, CODE.DUP
FLOAT	FLOAT.FROMBOOLEAN, FLOAT.FROMINTEGER, FLOAT.TAN, FLOAT.COS, FLOAT.SIN, FLOAT.>, FLOAT.<, FLOAT.%, FLOAT./, FLOAT.*, FLOAT.-, FLOAT.+, FLOAT.STACKDEPTH, FLOAT.SHOVE, FLOAT.YANKDUP, FLOAT.YANK, FLOAT.=, FLOAT.SWAP, FLOAT.POP, FLOAT.DUP
QGATE	QGATE.END, QGATE.MEASURE, QGATE.U2, QGATE.CPHASE, QGATE.SWP, QGATE.CNOT, QGATE.QNOT, QGATE.SRN, QGATE.U-THETA, QGATE.HADAMARD, QGATE.LIMITED-ORACLE, QGATE.GATE, QGATE.TRANSPOSE, QGATE.COMPOSE, QGATE.STACKDEPTH, QGATE.SHOVE, QGATE.YANKDUP, QGATE.YANK, QGATE.=, QGATE.SWAP, QGATE.POP, QGATE.DUP

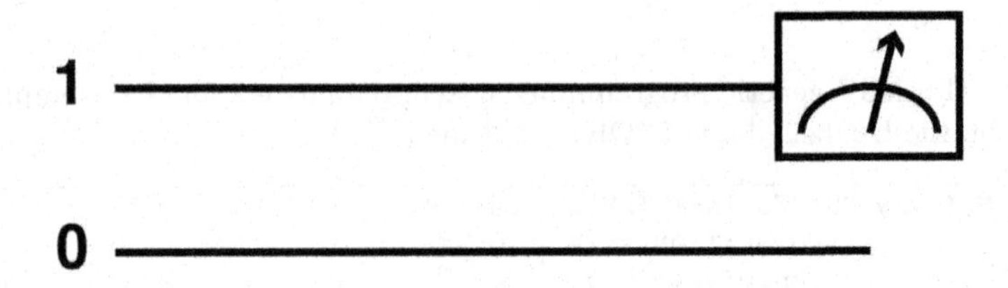

Figure 8.1. Gate array diagram for the empty "embryo" with which development begins for the solution to the Deutsch-Jozsa (XOR) problem. The only gate in the embryo performs a measurement of qubit 1; this need not even appear explicitly in the developed QGAME program as the call to `TEST-QUANTUM-PROGRAM` will specify that the final measurement will be performed on qubit 1. The developmental process will add gates from left to right, ending just before the measurement.

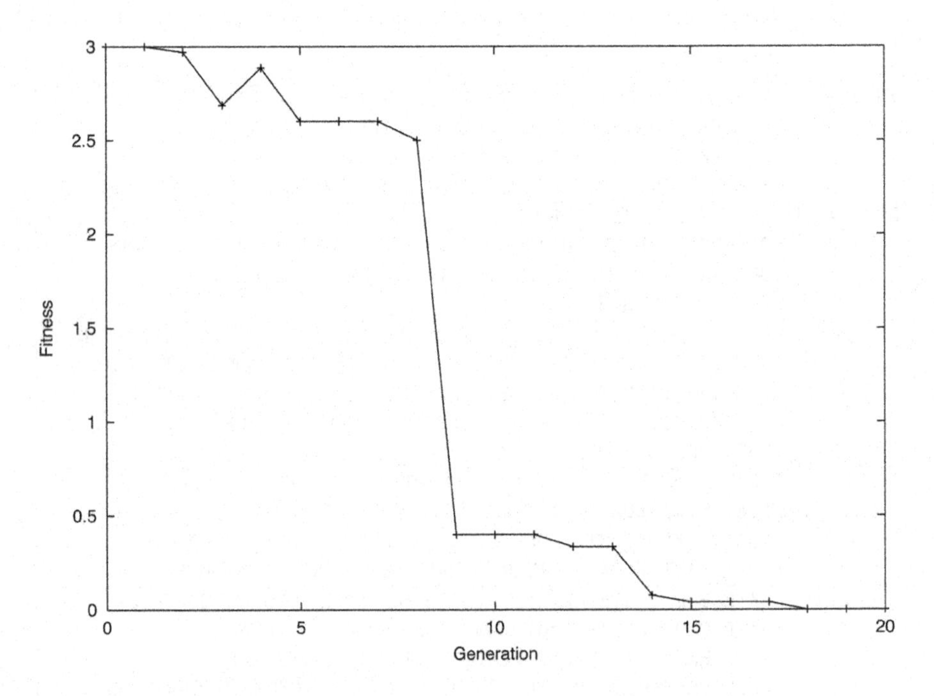

Figure 8.2. A plot of the fitnesses of the best individuals in each generation during a run of PushGP on the 1-bit Deutsch-Jozsa (XOR) problem.

Execution was aborted at generation 20, at which time the best reported program was as follows:

```
((BOOLEAN.= INTEGER.> CODE.DO*) ((FLOAT.TAN (FLOAT.<
(BOOLEAN.DUP (BOOLEAN.POP BOOLEAN.SHOVE INTEGER.-
QGATE.CPHASE (CODE.CAR CODE.LIST TRUE)))) (CODE.NULL
((CODE.APPEND) FLOAT.= (BOOLEAN.DUP BOOLEAN.DUP))))
CODE.CDR ((BOOLEAN.YANKDUP INTEGER.* BOOLEAN.=)
(0.16907119750976562D0) -2 (QGATE.SRN QGATE.STACKDEPTH
(QGATE.HADAMARD (QGATE.GATE CODE.STACKDEPTH)) CODE.NULL
(BOOLEAN.SWAP) (INTEGER.YANKDUP BOOLEAN.OR
(((QGATE.TRANSPOSE) CODE.NULL (QGATE.CPHASE INTEGER.>)
CODE.LIST) (QGATE.GATE ((-5 (FLOAT.STACKDEPTH)) CODE.YANK
BOOLEAN.POP))) (INTEGER.DUP)) QGATE.LIMITED-ORACLE))
(FLOAT.% QGATE.STACKDEPTH QGATE.GATE (((5 CODE.SWAP)
QGATE.LIMITED-ORACLE) FLOAT.YANK) FLOAT.SWAP FLOAT.TAN)
(TRUE)) (INTEGER.* (QGATE.SWP FLOAT.STACKDEPTH BOOLEAN.OR
CODE.CDR) BOOLEAN.STACKDEPTH))
```

Regardless of how this Push program is formatted, it is not clear from visual inspection how it works (and it has therefore been presented in the most economical format). Execution of this program produces, via development, the following QGAME program (as expressed in Lisp notation, where "#2A" indicates a 2-dimensional matrix, and with floating point numbers rounded to 4 decimal places):

```
((MATRIX-GATE #2A((0.7071 0.0 0.7071 0.0)
                  (0.0 0.7071 0.0 0.7071)
                  (0.7071 0.0 -0.7071 0.0)
                  (0.0 0.7071 0.0 -0.7071))
              ((HADAMARD 1)))
 (MATRIX-GATE #2A((0.7071 0.7071 0.0 0.0)
                  (-0.7071 0.7071 0.0 0.0)
                  (0.0 0.0 0.7071 0.7071)
                  (0.0 0.0 -0.7071 0.7071))
              (TRANSPOSED ((SRN 0))))
 (LIMITED-ORACLE 1 ORACLE-TT 1 0)
 (LIMITED-ORACLE 1 ORACLE-TT 0 1)
 (MATRIX-GATE #2A((0.7071 0.0 0.7071 0.0)
                  (0.0 0.7071 0.0 0.7071)
                  (0.7071 0.0 -0.7071 0.0)
                  (0.0 0.7071 0.0 -0.7071))
              ((HADAMARD 1))))
```

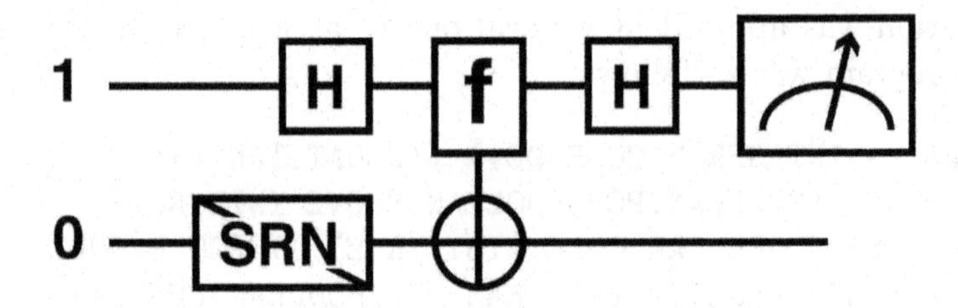

Figure 8.3. Gate array diagram for an evolved solution to the Deutsch-Jozsa (XOR) problem. The "f" gate is the oracle. The "SRN" gate with the diagonal line through it on qubit 0 transposed `Square Root of NOT` gate.

The second oracle call is redundant and can be removed; since the oracle limit is 1 a second call to `LIMITED-ORACLE` will have no effect. The first and final gates are simply `HADAMARD` gates applied to qubit 1, while the second gate is a transposed `SRN` ("square root of NOT"; see Chapter 2) gate. The final evolved, developed and simplified quantum program is diagrammed in Figure 8.3. This program solves the 1-bit version of the Deutsch-Jozsa (XOR) problem with 100% certainty using only a single oracle call.

How does this evolved solution solve the 1-bit Deutsch-Jozsa (XOR) problem? The mathematical explanation is straightforward — one needs only to construct and multiply all of the matrices — but it is difficult to provide an intuitive explanation even for such a simple quantum algorithm. The basic idea is indeed intuitive, however: the algorithm first puts both qubits into superpositions of $|0\rangle$ and $|1\rangle$ and then calls the oracle once on this superposition, extracting information about both classical inputs in a single call. This information must then be "decoded" from the resulting superposition by means of an additional `HADAMARD` gate, which reverses the effect of the `HADAMARD` gate prior to the oracle. Note that the final measurement is made on the qubit that is nominally the *input* to the oracle call, while the nominal output is ignored. This highlights one of the ways in which quantum gate arrays differ from classical logic circuits.[4] The oracle call in this case modifies qubit 0, but in doing so it changes every amplitude in the system state. Through this action (which is sometimes called the "back action" of a quantum gate) it changes the effect of the final `HADAMARD` on qubit 1, leading to the measurement of the correct answer for both possible inputs.

[4]The potential deceptiveness of quantum gate array diagrams that results from such differences was discussed in Chapter 3.

2. Grover's Database Search Problem

Grover's database search problem was described above in Chapters 1 and 2, the latter of which included a detailed presentation of one solution to the 4-item version of this problem. Grover's problem is an oracle problem, much like the Deutsch-Jozsa problem, except that the "promise" we are given regarding the oracle is different and the task is not just to distinguish two classes of oracles (uniform vs. balanced) but rather to determine exactly which of the possible oracles we have been given.

More specifically, we are promised, in the instance of the problem considered here, that the oracle will invert its output for one and only one input. Our task is to determine which input it is that produces the inversion. This is described as a database problem because we may think of the oracle as a database, for which all of the possible inputs are addresses, and we may think of the output inversion as an answer of "yes" to a database query for a marked item. Under this interpretation we are promised that we have been given a database containing a marked item at one and only one address, and we are asked to determine the address of that item using as few calls to the database query function (oracle) as possible. The number of queries required for a classical program to solve this problem with an n-item database is $n-1$ in the worst case, but Grover's algorithm can find the marked item in approximately $\sqrt{n}$ queries. For the 4-item database considered here Grover's algorithm requires only a single database query.

Techniques similar to those described above for the Deutsch-Jozsa problem also permit evolution of a solution to the 4-item database search problem.[5] Because the oracle is in this case a 3-qubit gate (two input qubits and one output qubit), one must use a quantum computer with at least 3 qubits. One must also designate two qubits for final measurements, rather than the one qubit required for Deutsch-Jozsa, since one must be able to read a 2-bit address (0, 1, 2, or 3) from the measurement qubits at the end of the simulation. The cases on which programs are tested for fitness are:

```
(((1 0 0 0) 0)
 ((0 1 0 0) 1)
 ((0 0 1 0) 2)
 ((0 0 0 1) 3))
```

[5]The evolution of a solution to this problem using using "stackless linear genome genetic programming," as described in Chapter 7, is documented in (Spector et al., 1999b).

Table 8.4. Push interpreter parameters for the example run of PushGP on the 4-item database search problem. Documentation on Push parameters and instructions is available from `http://hampshire.edu/lspector/push.html`.

MAX-RANDOM-FLOAT	10.0
MIN-RANDOM-FLOAT	-10.0
MAX-RANDOM-INTEGER	10
MIN-RANDOM-INTEGER	-10
EVALPUSH-LIMIT	250
MAX-POINTS-IN-RANDOM-EXPRESSIONS	50
MAX-POINTS-IN-PROGRAM	100
MAX-ORACLE-CALLS	1
Types	QGATE, FLOAT, CODE, INTEGER
Instructions	(see Table 8.6)

Table 8.5. PushGP genetic programming system parameters for the example run of PushGP on the 4-item database search problem.

MAX-NEW-POINTS-IN-MUTANTS	20
POPULATION-SIZE	25,000 ($\times$ 10 demes)
TOURNAMENT-SIZE	5
MUTATION-PROBABILITY	0.45
CROSSOVER-PROBABILITY	0.45
IMMIGRATION-PROBABILITY	0.005
MUTATION-OPERATORS	FAIR, GAUSSIAN-PERTURB, ADD, REMOVE
CROSSOVER-OPERATORS	STANDARD, FAIR
FITNESS-FUNCTION	10 $\times$ misses + max probability of error

This means that the answer, to be assembled from the measured values of two qubits (we'll specify these to be qubits 1 and 2, specifying the high-order and low-order bits of the answer respectively), should be 0 if the location of the marked item is (0, 0), 1 if the location is (0, 1), 2 if the location is (1, 0), and 3 if the location is (1, 1).

This problem was solved using PushGP with the parameters shown in Tables 8.4 and 8.5 and the instruction set shown in Table 8.6, running under the CMUCL open source Common Lisp system[6] on a 10-CPU cluster of 2.1GHZ Linux workstations. The complete source code for this run, along with the output logs, is available online.[7]

[6] `http://www.cons.org/cmucl/`

[7] `http://hampshire.edu/lspector/aqcp/evolved-grover/`

Table 8.6. Instructions used in the example run of PushGP on the 4-item database search problem.

INTEGER	INTEGER.FROMFLOAT, INTEGER./, INTEGER.*, INTEGER.-, INTEGER.+, INTEGER.SWAP, INTEGER.POP, INTEGER.DUP
CODE	CODE.DO*COUNT, CODE.DO*TIMES, CODE.FROMFLOAT, CODE.FROMINTEGER, CODE.DO, CODE.NTHCDR, CODE.NTH, CODE.APPEND, CODE.LIST, CODE.NOOP, CODE.IF, CODE.DO*, CODE.CONS, CODE.CDR, CODE.CAR, CODE.QUOTE, CODE.SWAP, CODE.POP, CODE.DUP
FLOAT	FLOAT.FROMINTEGER, FLOAT./, FLOAT.*, FLOAT.-, FLOAT.+, FLOAT.SWAP, FLOAT.POP, FLOAT.DUP
QGATE	QGATE.END, QGATE.MEASURE, QGATE.CPHASE, QGATE.SWP, QGATE.CNOT, QGATE.QNOT, QGATE.U-THETA, QGATE.HADAMARD, QGATE.LIMITED-ORACLE, QGATE.GATE, QGATE.TRANSPOSE, QGATE.COMPOSE, QGATE.SWAP, QGATE.POP, QGATE.DUP

The 10-CPU cluster was utilized by means of a scheme of "demes" like that described briefly in Chapter 4. PushGP was started on each of the nodes and the 10 runs were allowed to proceed asynchronously. After the fitness-testing step of each generation a pool of emigrants, consisting of 125 individuals (0.5% of the population size of 25,000) selected via fitness tournaments (with tournament size 5), was written to a shared file system, replacing any previous pool of emigrants from the same node. Following emigration, a randomly selected file of emigrants on the shared file system (which may have come from the same node or from a different node) is read and becomes the pool of immigrants from which the IMMIGRATION genetic operator will randomly select individuals in the next offspring-production step. If the attempt to read a file of emigrants from the shared file system fails for any reason (for example because of network problems) then the IMMIGRATION operator will act as a reproduction operator, producing clones of individuals from the current population.

This run also utilized the matrix literalization scheme discussed in Chapter 7. After the fitness-testing step of each generation the Push programs were processed in order of fitness (best first) until at least 10 matrix literals were obtained. This was accomplished by re-evaluating each Push program to produce, via development, a QGAME program, and by compressing strings of matrices in the developed QGAME program to produce compressed matrix literals. These literals were then available for inclusion in mutations performed during the next offspring-production step. In addition, this run utilized a GAUSSIAN-PERTURB

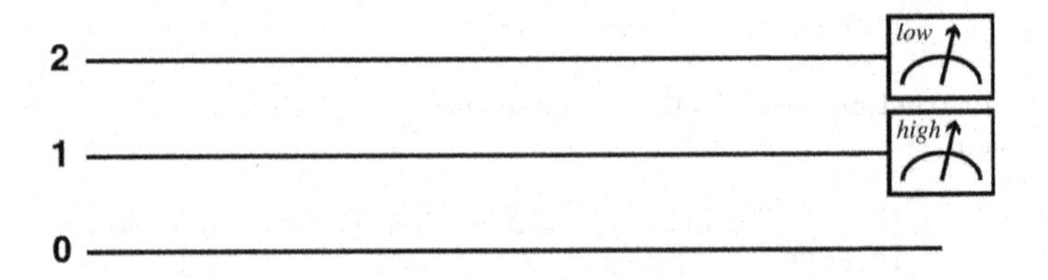

Figure 8.4. Gate array diagram for the empty "embryo" with which development begins for the solution to the database search problem. The only gates in the embryo perform measurement of qubits 1 (the high order bit of the answer) and 2 (the low order bit of the answer). The developmental process will add gates from left to right, ending just before the measurements.

genetic operator, the idea for which was described in Chapter 7. When this operator is chosen for a particular instance of mutation,[8] a child is produced from the parent by adding mean 0, standard deviation 0.01 Gaussian noise to each floating-point literal in the parent.

As with the Deutsch-Jozsa example in the previous section, the fitness of a Push program was assessed by running it once to produce a QGAME program (which began in this case with the empty "embryo" corresponding to the gate array shown in Figure 8.4), and by testing the QGAME program with the `TEST-QUANTUM-PROGRAM` function described in Chapter 2. The maximum permitted number of oracle calls per case was again 1, so that only the first oracle call in any developed QGAME program would have any effect. The fitness cases were those listed above and the threshold for a "miss" was again 0.48. Fitness was computed as the sum of 10 times the number of misses (the first return value from `TEST-QUANTUM-PROGRAM`) and the maximum probability of error for any one case (the second return value from `TEST-QUANTUM-PROGRAM`); this is the "lexicographic" fitness component combination scheme that was discussed in Chapter 7.

The fitnesses over the 10 demes are plotted in Figure 8.5. The elimination of "misses" is clearly visible as large drops in fitness values, which are lexicographic combinations of misses ($\times$ 10) and maximum probability of error per case. Fitness improvements within particular levels of misses are obscured by the scale, but Figure 8.6 shows the additional detail at the level of zero misses. The first deme to achieve a perfect fitness value of zero did so at generation 113, while the last deme to achieve a perfect fitness value did so at generation 152. The last of these

[8]In PushGP, a random one of the specified mutation operators is selected for each instance of mutation. Similarly for crossover: if multiple operators are specified then each instance of crossover uses a randomly selected crossover operator.

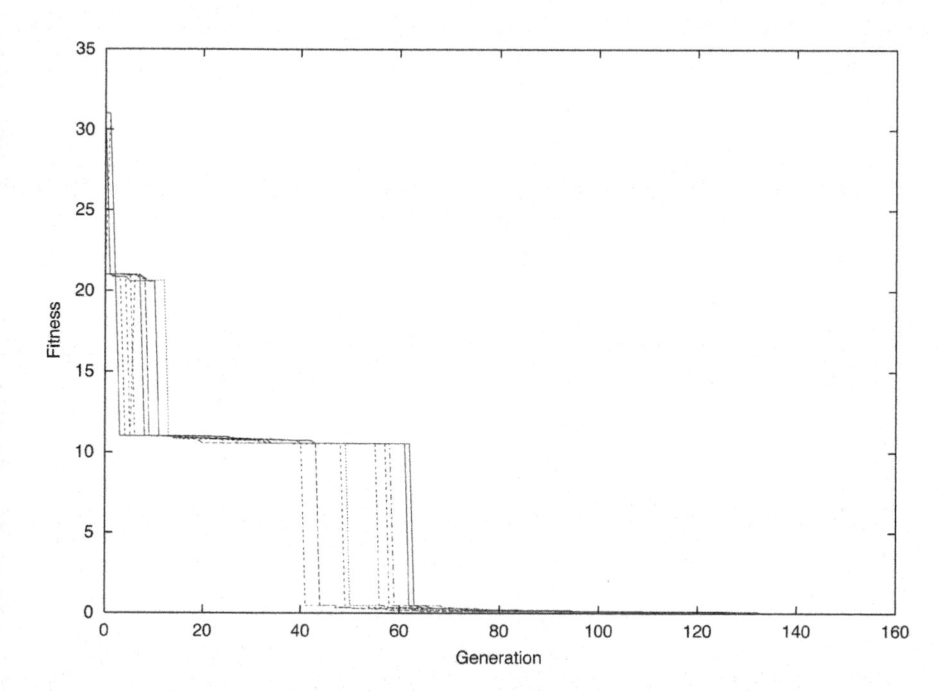

Figure 8.5. A plot of the fitnesses of the best individuals in each generation during a run of PushGP on the 4-item database search problem. This figure is dominated by the large drops due to the decreases in the "misses" component of the fitness function; it shows the overall structure of the evolutionary process but not the fine structure of fitness improvements at each level. Figure 8.6 shows a closer view of the improvements in fitness after all of the misses were eliminated. This run was conducted on a cluster of 10 computers that ran asynchronously, sharing individuals between generations (see text), and a line appears in the graph for each of the 10 runs. Because the individual runs ran asynchronously they reached particular generations at different times and one must be careful when inferring relations between runs from this graph; for example, an event that appears to the right of another event may actually have preceded that other event in time, and may even have influenced that other event via migration.

perfect-fitness individuals was chosen, arbitrarily, as the basis for the following analysis.

The evolved solution Push program contained 100 points, which was the maximum permitted.[9] The average number of points in the population that included this solution was 80.5, and the median fitness in this population was 0.0026. The solution Push program contained 5 unitary matrix literals, produced via the matrix literalization process described above, some of which were derived from other matrix literals earlier in

[9] Each instruction, literal, and pair of parentheses counts as one point.

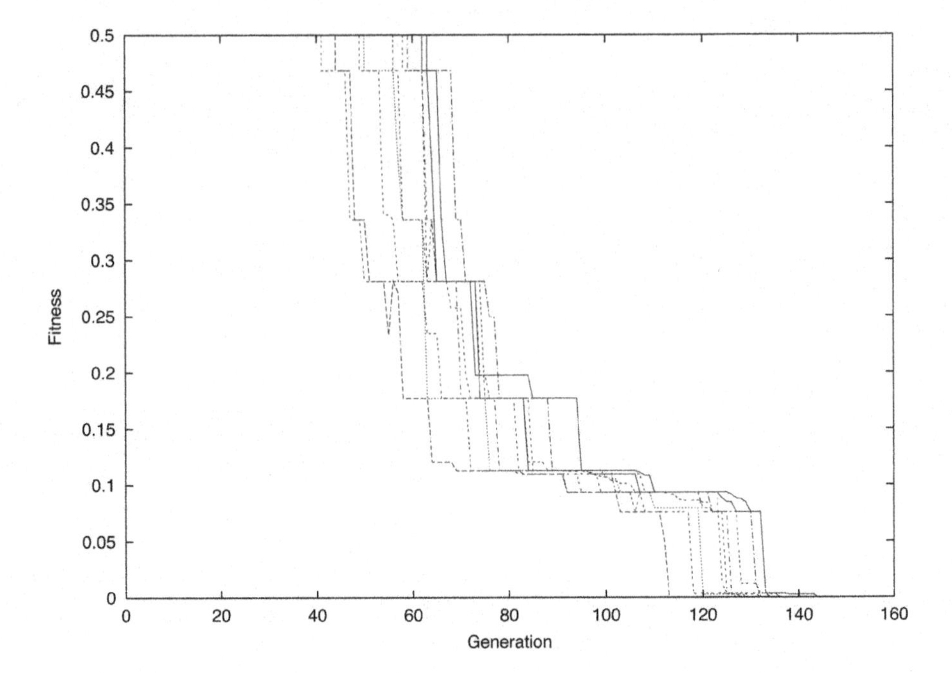

Figure 8.6. A plot of the fitnesses of the best individuals in each generation during a run of PushGP on the 4-item database search problem. This is a closer view of the graph in Figure 8.5, showing the improvements in fitness after all of the "misses" components of the fitness function were eliminated.

the evolutionary process. For example, one of the matrix literals is the composition of two instances of another matrix literal, which in turn includes three instances of a matrix that appears to have been produced by an earlier matrix literalization process. The inclusion of the matrix literals makes the printed representation of this Push program quite large (3,458 characters, not counting spaces); it is therefore not included here, although it can be found online.[10]

Execution of the evolved Push program produces, via development, a QGAME program consisting of 18 matrix gates. Some of the matrices in these gates appeared in the Push program as matrix literals, but others were produced by the execution of the Push program either from primitive gates or from matrix literals. For example, one matrix in the developed QGAME program is a transposed version of one of the matrix literals in the Push program. Another matrix in the developed QGAME program is a transposed version of one of the matrix literals in

[10]See `http://hampshire.edu/lspector/aqcp/evolved-grover/`, at the end of the log file `pushgp-output.n01.bw01.hampshire.edu`.

the Push program that has also been augmented by an additional `QNOT` gate. Again, because the textual version of this this program is verbose it is not included here.

As in the Deutsch-Jozsa example in the previous section, some of the gates in the final QGAME program are unnecessary and can be pruned from the result. Of particular interest in the present case is the fact that two of the gates, although they include matrix literals with rather complex histories, combine the matrices from those histories to produce identity operations; components of these histories are also used elsewhere in the final QGAME program to greater effect. The final QGAME program, after hand pruning and with the matrices removed for legibility, is as follows:

```
((HADAMARD 1)
 (MATRIX-GATE <matrix1> <history1>)
 (HADAMARD 1)
 (HADAMARD 0)
 (MATRIX-GATE <matrix2> <history2>)
 (LIMITED-ORACLE 1 ORACLE-TT 2 1 0)
 (HADAMARD 2)
 (MATRIX-GATE <matrix3> <history3>)
 (MATRIX-GATE <matrix4> <history4>)
 (HADAMARD 1))
```

The matrix indicated as `<matrix1>` is just a transposed version of the matrix indicated as `<matrix2>`, which has the following history:

```
((COMPRESSED
  ((COMPRESSED ((U-THETA 2 1.233552982796235)))
   (COMPRESSED
    ((COMPRESSED ((QNOT 0))) (COMPRESSED ((CNOT 1 2)))))))
```

The matrix indicated as `<matrix3>` has the following history:

```
((COMPRESSED ((HADAMARD 1)))
 (COMPRESSED
  ((COMPRESSED
    (TRANSPOSED ((U-THETA 1 1.0642909109545906))))))
 (COMPRESSED
  ((COMPRESSED
    (TRANSPOSED ((U-THETA 1 1.0642909109545906))))))
 (COMPRESSED
  ((COMPRESSED
    (TRANSPOSED ((U-THETA 1 1.0642909109545906)))))))
```

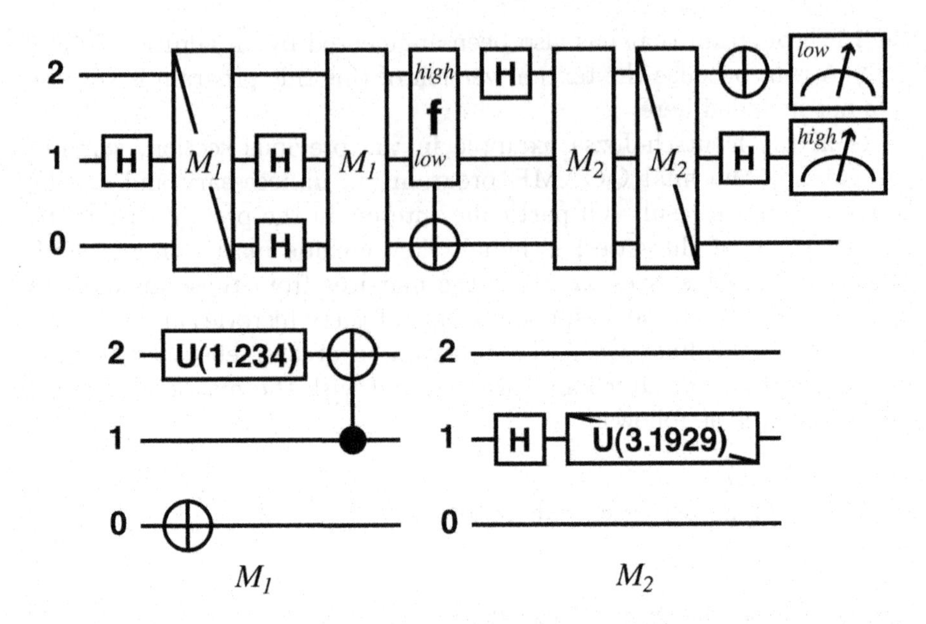

Figure 8.7. A gate array diagram for an evolved version of Grover's database search algorithm for a 4-item database. The full gate array is shown at the top, with M_1 and M_2 standing for the smaller gate arrays shown at the bottom. A diagonal line through a gate symbol indicates that the matrix for the gate is transposed. The "f" gate is the oracle.

The matrix indicated as `<matrix4>` is a transposed version of the matrix indicated as `<matrix3>`, to which a `QNOT` gate has also been added on qubit 2.

The resulting quantum gate array is diagrammed in Figure 8.7. M_1 in the figure corresponds to `<matrix2>` and M_2 corresponds to `<matrix3>`; the contents of each of these matrices are indicated in the smaller gate array diagrams in the bottom half of the figure. The transpositions in matrices 1 and 4 are indicated by the diagonal lines, and the additional `QNOT` gate that evolved as part of `<matrix3>` is drawn separately on the qubit 2 line in the main diagram. This gate array solves the 4-item database search problem with 100% certainty using only a single oracle call. The evolved gate array exhibits several forms of modularity, some of which were achieved via recursive matrix literalization and others of which owe to the code-manipulation and matrix-manipulation facilities of the Push instruction set used for this run.

How does this evolved solution work? At a general level of description the solution is the same as that presented in Section 3.3 above: a superposed state is fed into the call to the oracle gate and subsequent "decoding" gates extract the position of the marked item from the states

Table 8.7. Push interpreter parameters for the example run of PushGP on the Scaling Majority-ON problem. Documentation on Push parameters and instructions is available from `http://hampshire.edu/lspector/push.html`.

MAX-RANDOM-FLOAT	1.0
MIN-RANDOM-FLOAT	-1.0
MAX-RANDOM-INTEGER	10
MIN-RANDOM-INTEGER	-10
EVALPUSH-LIMIT	150
MAX-POINTS-IN-RANDOM-EXPRESSIONS	50
MAX-POINTS-IN-PROGRAM	100
MAX-ORACLE-CALLS	1
Types	QGATE, FLOAT, CODE, BOOLEAN, INTEGER
Instructions	(see Table 8.9)

in which the *address* qubits (as opposed to the *output* qubit) are left by the action of the oracle. The solution presented here is, however, considerably more complex than that presented in Section 3.3.[11] Part of the reason for this difference is that the result presented earlier was subjected to further human editing,[12] but part may also be due to an unfortunate evolutionary accident early in the run presented here. The oracle call in the evolved gate array uses qubit 2 as the high-order input and qubit 1 as the low-order input, while the measurements specified in the embryo use the opposite designation. If the programs that achieved limited success early in this run included the oracle call with this "backwards" configuration, then it may have been easier for evolution to find improvements that compensated for this configuration through additional gates than through the substitution of an alternative oracle configuration. Another factor contributing to the complexity of this solution may be the use of matrix literalization, which facilitates the evolution of quantum programs containing complex modules; while this probably extends the power of the automatic quantum computer programming system it may also have the unfortunate side effect of encouraging the generation of unnecessarily complex solutions.

[11]It is also considerably more complex than the solution evolved by the author previously using other techniques (Spector et al., 1999b).

[12]The editing performed here was limited to the removal of gates that had no effect on the result; further analysis may produce additional simplifications by substituting single gates for groups of gates, etc.

Table 8.8. PushGP genetic programming system parameters for the example run of PushGP on the Scaling Majority-ON problem.

MAX-NEW-POINTS-IN-MUTANTS	20
POPULATION-SIZE	5,000 (× 13 demes)
TOURNAMENT-SIZE	7
MUTATION-PROBABILITY	0.45
CROSSOVER-PROBABILITY	0.45
IMMIGRATION-PROBABILITY	0.005
MUTATION-OPERATORS	FAIR, PERTURB, ADD, REMOVE
CROSSOVER-OPERATORS	FAIR
FITNESS-FUNCTION	misses + max probability of error

3. Scaling Majority-ON

The Majority-ON problem asks us to determine if the number of inputs for which an oracle inverts its output qubit is greater than the number of inputs for which it does not invert its output qubit. The name comes from interpreting the inversion as an indication that the location in the oracle addressed by the input is "on," and asking whether a majority of the oracle's locations are "on" in this sense. For simplicity here we omit oracles with an equal number of "on" and "off" locations.

The Scaling Majority-ON problem presents the more ambitious task of evolving a scheme for solving Majority-ON that can be scaled up to work for an oracle of any size. To solve this problem using PushGP and QGAME a NUMQUBITS Push instruction was added that pushes the number of qubits in the current problem instance onto the INTEGER stack. For the fitness test the system was run using all permissible oracles with 1, 2, and 3 input qubits for the fitness test. For each fitness case with an n-input oracle the Push program was executed in the context of an embryo with $n+1$ qubits and a final measurement on the highest-numbered qubit. In addition, a global variable was set that caused the NUMQUBITS instruction to push n onto the INTEGER stack. This instruction could be used in evolved Push programs to alter the developmental process, thereby producing different QGAME programs for cases of different size.

Aside from the addition of the NUMQUBITS Push instruction, the techniques used to produce the Scaling Majority-ON program presented here were qualitatively similar to those used for the database search problem in the previous section. The full parameter and instruction sets are shown in Tables 8.7, 8.8 and 8.9. The complete source code for this run, along with the output logs, is available online.[13]

This run did not produce a completely successful solution, in the sense of "zero probability of error," although it did produce solutions that achieved zero misses. One such result, obtained at generation 112, was the following 94-point Push program:

```
((((INTEGER.SWAP NUMQUBITS INTEGER.*) NUMQUBITS INTEGER.*)
(((INTEGER.YANKDUP ((INTEGER.STACKDEPTH QGATE.HADAMARD)
CODE.DISCREPANCY (8)) (INTEGER.FROMFLOAT))
QGATE.LIMITED-ORACLE) ((CODE.YANK (INTEGER.STACKDEPTH))
(QGATE.DUP QGATE.U-THETA CODE.YANKDUP INTEGER.+))) (((
CODE.NTHCDR ((INTEGER.STACKDEPTH QGATE.HADAMARD)
CODE.DISCREPANCY)) BOOLEAN.=) (QGATE.GATE)) CODE.DO*
BOOLEAN.POP) NUMQUBITS (((((QGATE.GATE (CODE.STACKDEPTH
CODE.= (INTEGER.YANK CODE.NTH CODE.STACKDEPTH (FLOAT.POP
(FLOAT.STACKDEPTH))) CODE.APPEND) (FLOAT.TAN) ((
BOOLEAN.FROMFLOAT)))) (BOOLEAN.=) FLOAT.STACKDEPTH ((
NUMQUBITS (CODE.DO*TIMES)) (((QGATE.QNOT (
-0.25270235538482666d0)) NUMQUBITS) QGATE.LIMITED-ORACLE))
NIL CODE.IF)) (QGATE.U-THETA QGATE.DUP QGATE.GATE))
QGATE.CPHASE)
```

For the 1-input fitness cases this program produces the following QGAME program:[14]

```
((LIMITED-ORACLE 1 ORACLE-TT 0 1)
 (HADAMARD 0)
 (LIMITED-ORACLE 1 ORACLE-TT 0 1)
 (HADAMARD 1)
 (HADAMARD 1)
 (LIMITED-ORACLE 1 ORACLE-TT 0 1)
 (U-THETA 0 6.03048295179476)
 (U-THETA 0 6.03048295179476))
```

[13] `http://hampshire.edu/lspector/aqcp/evolved-majon/`

[14] In this and subsequent listings any `MATRIX-GATE`s with histories containing only a single primitive gate are replaced by the primitive gates themselves for readability. In this particular run a low limit of 5 on history nesting depth — see page 79 — prevented the production of non-trivial `MATRIX-GATE`s.

For the 2-input fitness cases it produces the following QGAME program:

```
((HADAMARD 0)
 (HADAMARD 1)
 (LIMITED-ORACLE 1 ORACLE-TT 0 1 2)
 (U-THETA 0 6.03048295179476)
 (U-THETA 0 6.03048295179476)
 (LIMITED-ORACLE 1 ORACLE-TT 0 1 2)
 (U-THETA 0 6.03048295179476))
```

For the 3-input fitness cases it produces the following QGAME program:

```
((HADAMARD 0)
 (HADAMARD 2)
 (HADAMARD 1)
 (LIMITED-ORACLE 1 ORACLE-TT 0 1 2 3)
 (U-THETA 0 6.03048295179476)
 (U-THETA 0 6.03048295179476))
```

Many of the gates in the first two of these programs are superfluous; those that are not are diagrammed in Figure 8.8. These quantum programs, which are similar to those evolved earlier with somewhat simpler techniques (Spector et al., 1999b), do indeed solve the Majority-ON problem with a maximum probability of error less than 50% for oracles of all sizes. Although the evolved Push program does not scale properly to oracles larger than those used in the fitness test — that is, running the Push program in the context of a larger embryo and a larger value for NUMQUBITS does not produce an appropriate QGAME program for the larger oracle — it is clear from visual inspection how this algorithm can be scaled up indefinitely. Unfortunately, however, the maximum probabilities of error for the cases shown in the Figure are 0, 0.25, and 0.375, and the probabilities continue to approach 50% very quickly as the oracle sizes increase. In fact, these solutions are equivalent to the simple probabilistic classical algorithm of querying a single, random location of the oracle and answering "yes" if and only if the corresponding location of the oracle is "on." But although the evolved programs are not better than classical in this case, the example nonetheless demonstrates how genetic programming can be used as an aid in the development of scalable quantum algorithms.

Table 8.9. Instructions used in the example run of PushGP on the Scaling Majority-ON problem.

INTEGER	NUMQUBITS, INTEGER.FROMBOOLEAN, INTEGER.FROMFLOAT, INTEGER.>, INTEGER.<, INTEGER.%, INTEGER./, INTEGER.*, INTEGER.-, INTEGER.+, INTEGER.STACKDEPTH, INTEGER.SHOVE, INTEGER.YANKDUP, INTEGER.YANK, INTEGER.=, INTEGER.SWAP, INTEGER.POP, INTEGER.DUP
BOOLEAN	BOOLEAN.FROMFLOAT, BOOLEAN.FROMINTEGER, BOOLEAN.NOT, BOOLEAN.OR, BOOLEAN.AND, BOOLEAN.STACKDEPTH, BOOLEAN.SHOVE, BOOLEAN.YANKDUP, BOOLEAN.YANK, BOOLEAN.=, BOOLEAN.SWAP, BOOLEAN.POP, BOOLEAN.DUP
CODE	CODE.DO*COUNT, CODE.DO*TIMES, CODE.FROMBOOLEAN, CODE.FROMFLOAT, CODE.FROMINTEGER, CODE.DISCREPANCY, CODE.DO, CODE.NTHCDR, CODE.NTH, CODE.APPEND, CODE.LIST, CODE.NOOP, CODE.IF, CODE.DO*, CODE.CONS, CODE.CDR, CODE.CAR, CODE.NULL, CODE.ATOM, CODE.QUOTE, CODE.STACKDEPTH, CODE.SHOVE, CODE.YANKDUP, CODE.YANK, CODE.=, CODE.SWAP, CODE.POP, CODE.DUP
FLOAT	FLOAT.FROMBOOLEAN, FLOAT.FROMINTEGER, FLOAT.TAN, FLOAT.COS, FLOAT.SIN, FLOAT.>, FLOAT.<, FLOAT.%, FLOAT./, FLOAT.*, FLOAT.-, FLOAT.+, FLOAT.STACKDEPTH, FLOAT.SHOVE, FLOAT.YANKDUP, FLOAT.YANK, FLOAT.=, FLOAT.SWAP, FLOAT.POP, FLOAT.DUP
QGATE	QGATE.END, QGATE.MEASURE, QGATE.U2, QGATE.CPHASE, QGATE.SWP, QGATE.CNOT, QGATE.QNOT, QGATE.SRN, QGATE.U-THETA, QGATE.HADAMARD, QGATE.LIMITED-ORACLE, QGATE.GATE, QGATE.TRANSPOSE, QGATE.COMPOSE, QGATE.STACKDEPTH, QGATE.SHOVE, QGATE.YANKDUP, QGATE.YANK, QGATE.=, QGATE.SWAP, QGATE.POP, QGATE.DUP

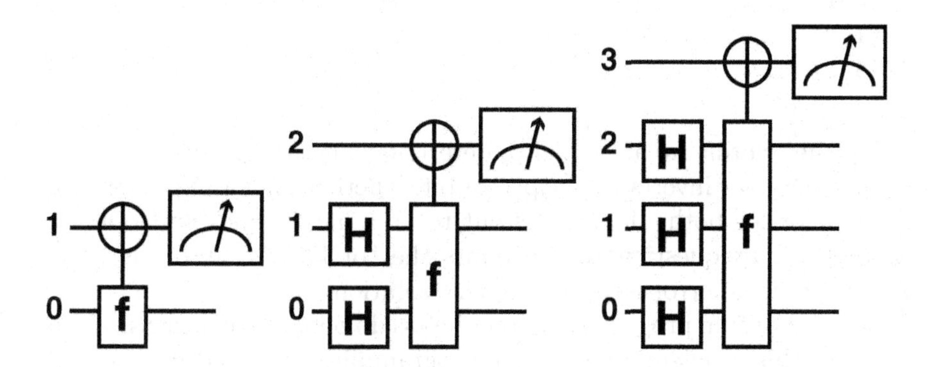

Figure 8.8. Gate array diagrams for the "Majority ON" problem for various oracle sizes, produced by a genetic programming run that evolved scalable programs. These are not better than classical solutions.

Table 8.10. Push interpreter parameters for the example runs of PushGP on the OR and AND/OR problems. Documentation on Push parameters and instructions is available from **http://hampshire.edu/lspector/push.html**.

MAX-RANDOM-FLOAT	1.0
MIN-RANDOM-FLOAT	-1.0
MAX-RANDOM-INTEGER	9
MIN-RANDOM-INTEGER	-10
EVALPUSH-LIMIT	150
MAX-POINTS-IN-RANDOM-EXPRESSIONS	50
MAX-POINTS-IN-PROGRAM	100
MAX-ORACLE-CALLS	1
Types	QGATE, FLOAT, CODE, INTEGER
Instructions	(see Table 8.12)

4. The OR and AND/OR Problems

The OR and AND/OR problems are oracle problems similar to the XOR problem described above, but they ask us to determine a different property of the oracles. The OR problem is identical to the XOR problem except that we are asked to determine the truth of the logical formula $I_0 \vee I_1$, where I_0 means "inverts with input 0," I_1 means "inverts with input 1," and $\vee$ is the (inclusive) OR function. In the notation used for QGAME's TEST-QUANTUM-PROGRAM function, the cases that we use to assess fitness are:

```
(((0 0) 0)
 ((0 1) 1)
 ((1 0) 1)
 ((1 1) 1))
```

In other words, we are asked to determine whether the oracle we have been given *ever* inverts its output qubit, whether for a 0 input, or for a 1 input, or for both. This turns out to be a harder question to answer than the XOR question (which omits the "or both"), and it is known that there is no error-free single query solution.

But a quantum program can nonetheless do better than a classical program on this problem, and genetic programming was used to discover a quantum algorithm that performed better than any that had previously been published. The evolved quantum program has a maximum probability of error of $\frac{1}{10}$. This is better than can be achieved using even a probabilistic classical program, which must necessarily have a max-

Table 8.11. PushGP genetic programming system parameters for the example runs of PushGP on the OR and AND/OR problems.

MAX-NEW-POINTS-IN-MUTANTS	10
POPULATION-SIZE	50,000 ($\times$ 13 demes)
TOURNAMENT-SIZE	7
MUTATION-PROBABILITY	0.48
CROSSOVER-PROBABILITY	0.48
IMMIGRATION-PROBABILITY	0.005
MUTATION-OPERATORS	PERTURB, ADD, REMOVE
CROSSOVER-OPERATORS	FAIR
SIZE-PRESSURE	2, IDEAL-SIZE= 50
FITNESS-FUNCTION	if misses = 0 then: $0.1 \times p_{max}$ otherwise: $(0.1 \times p_{max}) + \left\lfloor 10^6 \times \frac{\sum_{i=1}^{n} \frac{1}{1+e^{-\psi(p_i - 0.48)}}}{n} \right\rfloor$ where: n = number of fitness cases, p_i = probability of error for case i, p_{max} = maximum probability of error, and $\psi = e^{(e+1)}$

imum probability of error of at least $\frac{1}{6}$. The evolved program, which was originally produced using the LGP genetic programming system[15] and a precursor to QGAME, is presented along with an analysis of the problem's classical and quantum complexity in (Spector et al., 1999a) and (Barnum et al., 2000).

In this section we describe the more recent evolution of an equivalent quantum algorithm using PushGP and QGAME. For this run an alternative, stackless implementation of the QGATE data type was used. There was no QGATE.GATE Push instruction and the execution of Push instructions corresponding to primitive quantum gates (such as QGATE.HADAMARD) sent QGAME instructions directly to the developing embryo. This decreased the amount of Push code required to build simple QGAME programs, but it did not allow the Push program to manipulate and store novel unitary matrices during development.

The implementation of QGATE.MEASURE in this run was also unusual. The implementation used in the previous examples simply added an instruction expression, "(measure q)," to the developing embryo, with q taken from the INTEGER stack (modulo the number of qubits in the sys-

[15] Available from http://helios.hampshire.edu/lspector/code.html.

Table 8.12. Instructions used in the example runs of PushGP on the OR and AND/OR problems. These runs used alternative implementations of the QGATE instructions (see text).

INTEGER	INTEGER.MAX, INTEGER.MIN, INTEGER.%, INTEGER./, INTEGER.*, INTEGER.-, INTEGER.+, INTEGER.STACKDEPTH, INTEGER.SHOVE, INTEGER.YANKDUP, INTEGER.YANK, INTEGER.SWAP, INTEGER.POP, INTEGER.DUP
CODE	CODE.QUOTE, CODE.SWAP, CODE.POP, CODE.DUP
FLOAT	FLOAT.TAN, FLOAT.COS, FLOAT.SIN, FLOAT.MAX, FLOAT.MIN, FLOAT.%, FLOAT./, FLOAT.*, FLOAT.-, FLOAT.+, FLOAT.STACKDEPTH, FLOAT.SHOVE, FLOAT.YANKDUP, FLOAT.YANK, FLOAT.SWAP, FLOAT.POP, FLOAT.DUP
QGATE	QGATE.MEASURE, QGATE.HALT, QGATE.U2, QGATE.CPHASE, QGATE.SWP, QGATE.CNOT, QGATE.QNOT, QGATE.SRN, QGATE.U-THETA, QGATE.HADAMARD, QGATE.LIMITED-ORACLE

tem). Subsequent calls to QGATE.END were required to complete the branches of the computation for the two possible measurement outcomes (0 and 1).[16] For the present run an alternative implementation of QGATE.MEASURE was used that ensures, assuming that the Push program that contains it runs to completion, that all measurements are followed by complete branches for both possible outcomes. QGATE.MEASURE does this by taking two arguments from the CODE stack in addition to the index of the qubit to be measured (which is taken from the INTEGER stack). It then does the following:

- Adds the MEASURE expression to the developing QGAME program.
- Recursively executes one of the popped pieces of code (the one that was deeper in the stack), possibly adding additional elements to the developing QGAME program in the process.
- Adds an (END) to the developing QGAME program.
- Recursively executes the other popped piece of code, possibly adding additional elements to the developing QGAME program.
- Adds another (END) to the developing QGAME program.

The other parameters for this run are shown in Tables 8.10, 8.11, and 8.12. The SIZE-PRESSURE parameter referred to in Table 8.10 relates to an experimental feature of PushGP that is intended to help control

[16]See page 26 for the syntax of measurement constructions in QGAME.

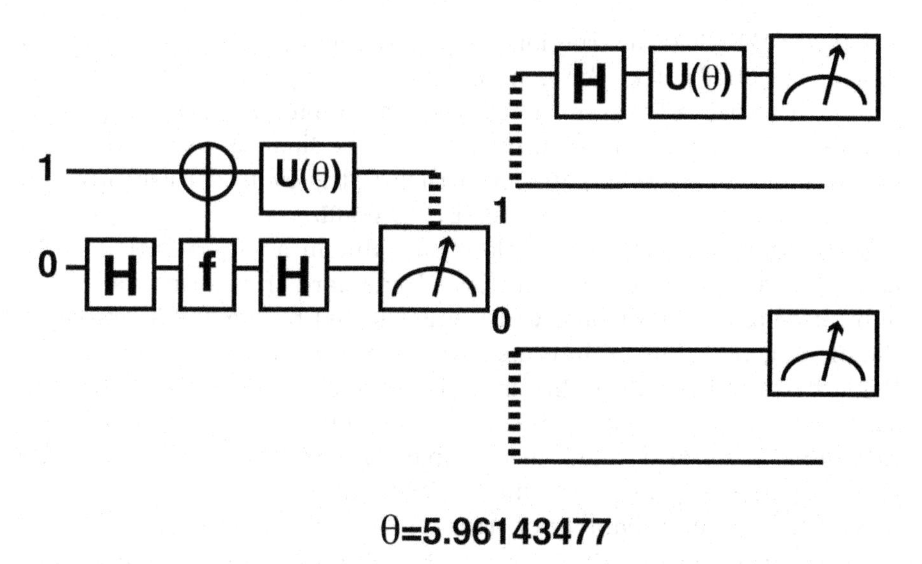

Figure 8.9. A gate array diagram for an evolved solution to the OR oracle problem. The gate marked "f" is the oracle. The two sub-diagrams on the right represent the two possible execution paths following the intermediate measurement. In the bottom sub-diagram the result of the intermediate measurement is 0 and the result of the overall computation is read immediately from the other qubit. In the top sub-diagram the result of the intermediate measurement is 1 and additional gates are applied to the other qubit prior to the final measurement.

program bloat; when this feature is enabled each attempt to use a genetic operator causes the operator to be called the indicated number of times (2 in this case), producing that number of potential offspring. The single offspring closest in size to the specified `IDEAL-SIZE` is chosen from these, and the others are discarded.

The fitness function for programs that achieve zero misses is the maximum probability of error on any single fitness case times 0.1. For programs with misses, however, the fitness function is a lexicographic combination of a sigmoid function (based on the differences between each probability of error and the "miss threshold") and the maximum probability of error. As discussed in Chapter 7, this sigmoid function provides a smoother fitness landscape while still prioritizing the elimination of misses, although the effectiveness of this measure has not been empirically tested.

The gate array in Figure 8.9 shows one result of this run, obtained at generation 302 and simplified by hand. This result exhibits elements of modularity even though it used only a minimal subset of Push's code-manipulation instructions and only one instruction — the modified QGATE.MEASURE instruction — that triggers recursive execution of code on the CODE stack. For example, the same angle appears twice as an argument to U-THETA, even though there are no duplicate floating point literals in the evolved Push program, and the final QGAME program includes three HADAMARD gates even though the evolved Push program contains only two instances of QGATE.HADAMARD.

This algorithm calls the oracle on a qubit in a superposition of $|0\rangle$ and $|1\rangle$ and then, after an additional Hadamard transformation of the qubit used as the input (and which was affected by the "back action" of the oracle), performs an intermediate measurement of the input qubit. Regardless of the result of this intermediate measurement, the final measurement is made on qubit 1 (as was specified in the embryo), but in one case qubit 1 is transformed, using copies of gates that appeared earlier in the algorithm, prior to the final measurement.

The maximum probability of error for this algorithm is $\frac{1}{10}$, while classical algorithms necessarily have a probability of error of at least $\frac{1}{6}$. The existence of quantum algorithms with a maximum probability of error of $\frac{1}{10}$ was first discovered by genetic programming.

The AND/OR problem extends the OR problem to a larger oracle and to a more complex logical property. In this problem we are asked to determine if the cases for which the 2-qubit oracle flips its output qubit satisfy the logical formula $(I_{00} \vee I_{01}) \wedge (I_{10} \vee I_{11})$, where $\wedge$ is the AND function. This formula is illustrated as an "and/or tree" in Figure 8.10. In the notation used for QGAME's TEST-QUANTUM-PROGRAM function, the cases that we use to assess fitness are:

```
(((0 0 0 0) 0)
 ((0 0 0 1) 0)
 ((0 0 1 0) 0)
 ((0 0 1 1) 0)
 ((0 1 0 0) 0)
 ((0 1 0 1) 1)
 ((0 1 1 0) 1)
 ((0 1 1 1) 1)
 ((1 0 0 0) 0)
 ((1 0 0 1) 1)
 ((1 0 1 0) 1)
 ((1 0 1 1) 1)
 ((1 1 0 0) 0)
 ((1 1 0 1) 1)
 ((1 1 1 0) 1)
 ((1 1 1 1) 1))
```

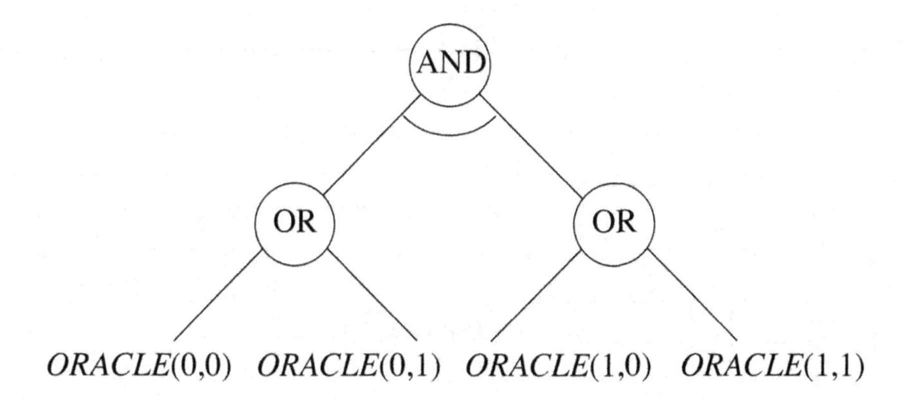

Figure 8.10. An AND/OR tree describing the nature of the AND/OR oracle problem.

The existence of better-than-classical quantum algorithms for the AND/OR problem was first discovered by genetic programming. The first evolved programs for this problem (which were also evolved using LGP and a predecessor to QGAME) are presented, along with a complexity analysis, in (Spector et al., 1999a) and (Barnum et al., 2000). Here we present a program equivalent to the best of these that was evolved more recently using PushGP and QGAME, with the same parameters as those used for the run on the OR problem above (Tables 8.10, 8.11, and 8.12); only the fitness cases and the size of the embryo were changed.

The evolved quantum program, a hand-simplified version of which is shown in Figure 8.11, has a maximum probability of error of 0.28731. By contrast the best that can be achieved by a probabilistic classical program is an error probability of $\frac{1}{3}$. Like the solution to the OR problem above, this algorithm works by calling the oracle on inputs in superposition and by subsequently performing intermediate measurements on the input qubits, which will have been affected by the back action of the oracle call. The final measurement is again made on the oracle's output qubit, but only after additional transformations to the output qubit that are conditional on the intermediate measurements.

It is also noteworthy that the Push program that produced this solution contained only one instance of `QGATE.MEASURE`, meaning that the multiple-measurement solution resulted from the use of the use of Push's code-manipulation instructions, only a minimal subset of which were included in this run.

It is natural to ask how these algorithms, both for the OR problem and for the AND/OR problem, can be scaled up to larger problem in-

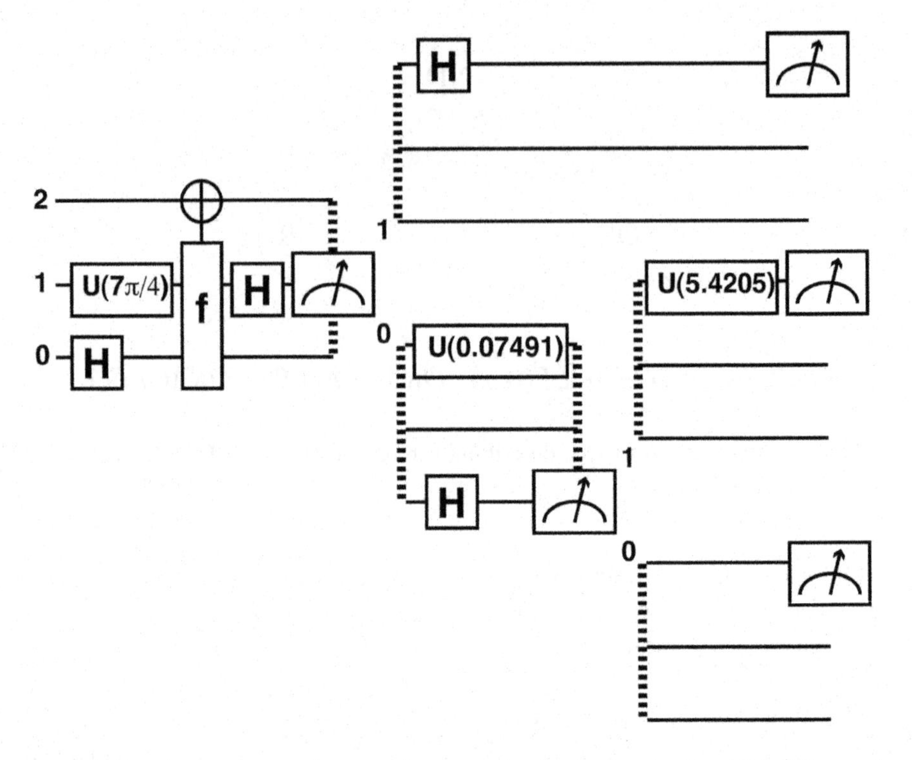

Figure 8.11. A gate array diagram for an evolved solution to the AND/OR oracle problem. The gate marked "f" is the oracle. The sub-diagrams on the right represent the possible execution paths following the intermediate measurements.

stances. Unfortunately, simple concatenations of the evolved algorithms do not suffice for this purpose. It is possible, however, that solutions to larger problem instances may be discovered through future genetic programming runs, and that the principles by which these algorithms can be scaled up can subsequently be inferred.

5. Gate Communication Problems

This section describes several problems that emerged from explorations of the relations between the communication and entanglement-generation capacities of certain quantum gates (Spector and Bernstein, 2003; Bennett et al., 2004). These explorations involved several iterative cycles of problem formulation, genetic programming, and human analysis. All of the genetic programming runs used PushGP, QGAME, and techniques similar to those described above. Due to space limitations the details of the many individual runs will not be presented here, except for the few novel features introduced specifically for these problems.

In the course of this work John Smolin defined the following gate, which was initially thought to generate entanglement without allowing for communication:

$$SMOLIN \equiv \begin{bmatrix} \frac{1}{\sqrt{2}} & 0 & 0 & \frac{1}{\sqrt{2}} \\ 0 & 1 & 0 & 0 \\ 0 & 0 & 1 & 0 \\ \frac{1}{\sqrt{2}} & 0 & 0 & -\frac{1}{\sqrt{2}} \end{bmatrix}$$

The open question was whether two parties (Alice and Bob) who were allowed to interact with one another only through a single use of this gate could use that interaction to communicate. This problem was solved using PushGP and QGAME, along with a developmental restriction that prevented any gates, aside form a single instance of `SMOLIN`, from spanning Alice's and Bob's qubits. The developmental restriction was implemented in the code that adds a gate expression to the developing embryo: if the gate expression spanned Alice's and Bob's qubits then it was simply ignored, unless it was both a `SMOLIN` gate expression and the first such expression encountered in the developmental process.

There are two fitness cases in this problem. In the first case we leave Alice's qubit in the 0 state and penalize a program for any probability of reading a 1 from Bob's qubit at the end of the computation. In the second case we initially invert Alice's qubit and we penalize a program for any probability of reading a 0 from Bob's qubit at the end of the computation. Ideally Bob's qubit will always be read to have the value at which we initially set Alice's qubit. This fitness test can be implemented using techniques similar to those discussed above, using a 0-input `ORACLE` gate (which will act either as an identity transformation or as an uncontrolled `QNOT` to implement Alice's choice). A call to this oracle is included, on Alice's qubit, at the beginning of the embryo; the answer read from Bob's qubit at the end of the computation should be 0 when the oracle is the identity transformation and 0 when it is a `QNOT`.

The evolved and hand-simplified quantum program shown in Figure 8.12 solved this problem by determining, unexpectedly, that a single classical bit can be communicated through a single application of the `SMOLIN` gate with zero probability of error. This was a useful contribution to the human discovery (by Herbert J. Bernstein) of a general strategy for communicating, without any probability of error, through a generalization of the `SMOLIN` gate called $J(\theta)$:

$$J(\theta) \equiv \begin{bmatrix} \cos(\theta) & 0 & 0 & \sin(\theta) \\ 0 & 1 & 0 & 0 \\ 0 & 0 & 1 & 0 \\ \sin(\theta) & 0 & 0 & -\cos(\theta) \end{bmatrix}$$

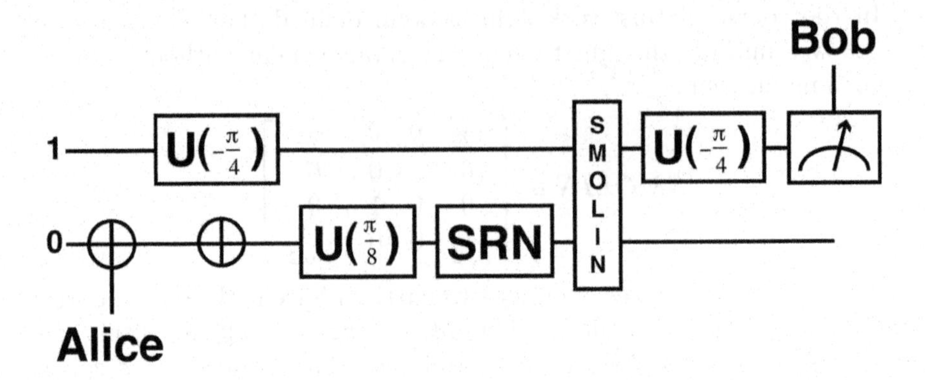

Figure 8.12. A gate array diagram for an evolved protocol for communicating one classical bit through a Smolin gate. Alice either does or does not flip qubit 0 to send a 0 or a 1, respectively; the gate that sets her message is part of the embryo in the developmental process driven by the evolved PushGP program.

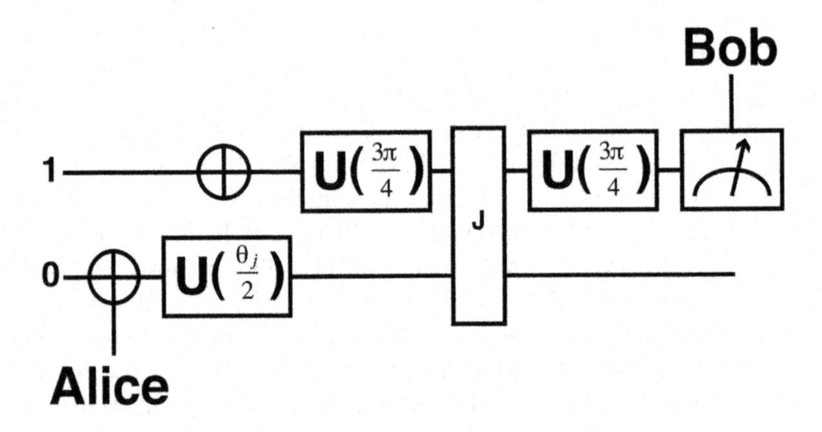

Figure 8.13. A gate array diagram for an evolved protocol for communicating one classical bit through a $J(\theta)$ gate.

The strategy for communicating through $J(\theta)$ shown in Figure 8.13 was designed by hand, but it was derived largely from the evolved strategy for communicating through SMOLIN shown in Figure 8.12. One interesting feature of this strategy for communicating through $J(\theta)$ is that Bob does not even need to know the angle θ used in the $J(\theta)$ gate in order to decode Alice's bit; Alice must know θ in order to apply the appropriate rotation to her qubit prior to the application of $J(\theta)$, but Bob can perform the same decoding steps regardless of θ.

Further analysis of the SMOLIN and $J(\theta)$ communication strategies yielded new problems to which genetic programming was subsequently applied. In particular, it led to the definition of the following $BS(\theta)$ (Bernstein Spector) gate:

$$BS(\theta) \equiv \begin{bmatrix} \cos(\theta) & 0 & 0 & \sin(\theta) \\ 0 & 0 & 1 & 0 \\ 0 & 1 & 0 & 0 \\ \sin(\theta) & 0 & 0 & -\cos(\theta) \end{bmatrix}$$

At the time of this writing the $BS(\theta)$ gate appears to entangle more than it can communicate, and communication appears difficult except at $\theta \bmod \pi = 0$ (Spector and Bernstein, 2003; Bennett et al., 2004). Genetic programming has been used to explore several questions related to this gate, including the communication capacity that it provides for various values of θ. In some cases the techniques described in Section 8.3 above, for evolving scalable quantum algorithms, were used with the modification that θ, rather than the number of qubits, was varied between fitness cases.

One example result, shown in Figure 8.14, involves communication in the context of prior entanglement. We stipulate that Alice and Bob, prior to the time at which communication via the $BS(\theta)$ gate is required, entangle two of their qubits. In the genetic programming run we create this prior entanglement by including a HADAMARD gate and a CNOT gate in the "embryo" from which the QGAME program develops. The genetic programming result shown in Figure 8.14 demonstrates that it is possible, in the context of prior entanglement, for Alice to send Bob a classical bit through $BS(\frac{\pi}{4})$ with no probability of error. The algorithm for doing this, as shown in the figure, is extremely simple; aside from the elements that were included in the embryo only two HADAMARD gates and the call to BS-THETA itself are required.

A related result is shown in Figure 8.15. In the run that produced this result we included code to generate prior entanglement in the embryo, as above, but we then attempted to transmit *two* classical bits through a single application of $BS(\pi)$. Genetic programming found a way to do this, with no probability of error, using the quantum program shown in the figure. This result is a form of a well-known phenomenon called "superdense coding," in which two bits of classical information can be transmitted through a single qubit channel.

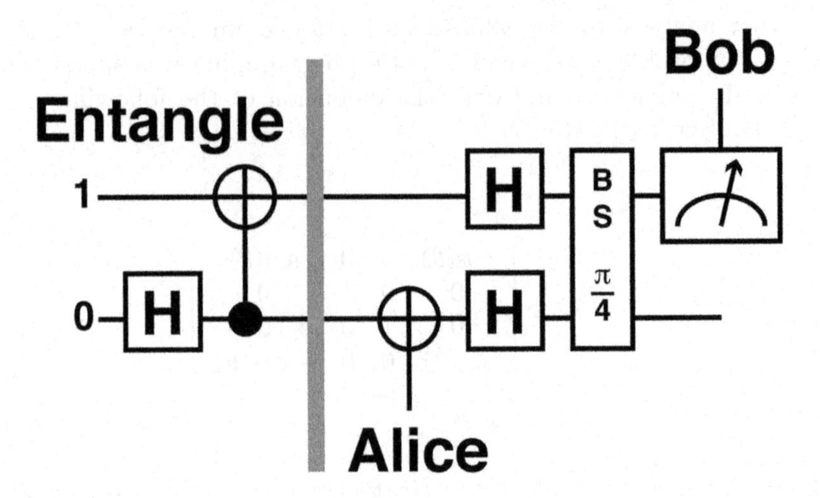

Figure 8.14. A gate array diagram for an evolved protocol for communicating one classical bit through a $BS(\frac{\pi}{4})$ gate in the context of prior entanglement. The entanglement-generating gates, to the left of the vertical bar, were included in the embryo to which the developmental process was applied.

6. Significance of These Results

Most of the results presented in this chapter demonstrate the human competitive nature of genetic and evolutionary computing technologies. A few also demonstrate the production, via genetic programming, of genuinely new knowledge with respect to the nature and power of quantum computing.

What is meant by "human competitive" in this context? John Koza and his colleagues have developed a list of eight criteria for the assertion of human competitiveness of results produced by intelligent technologies (Koza et al., 2003). These criteria are expressed relative to measures that are commonly employed to assess *human* contributions to scientific and technological research and development, such as patents and publications in reputable, peer-reviewed scientific journals. The criteria all focus on properties of the results themselves, not on their automatic production by computer systems.

Several of Koza's criteria apply to the results presented in this chapter. Two that are particularly helpful in assessing the significance of these results are the following:

- B: The result is equal to or better than a result that was accepted as a new scientific result at the time when it was published in a peer-reviewed scientific journal.

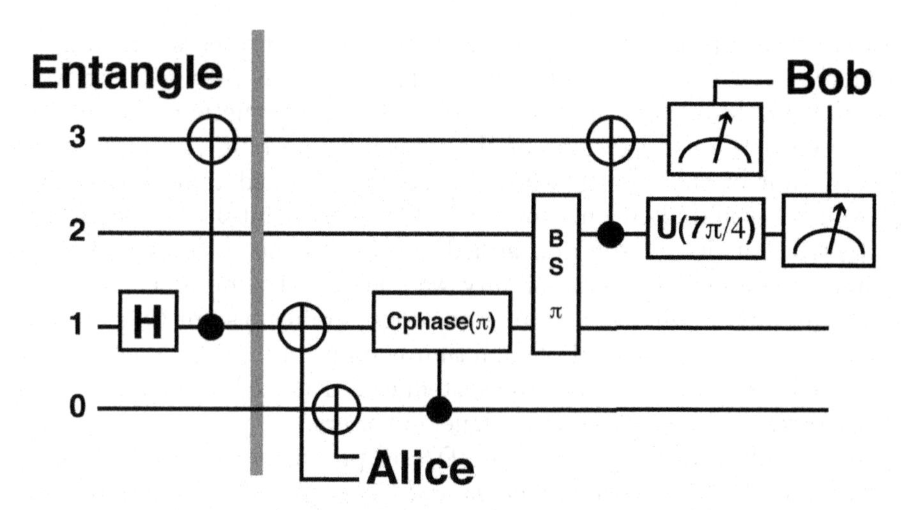

Figure 8.15. A gate array diagram for an evolved protocol for communicating two classical bits through one application of a $BS(\pi)$ gate in the context of prior entanglement. This is a form of quantum superdense coding re-discovered by genetic programming. The entanglement-generating gates, to the left of the vertical bar, were included in the embryo to which the developmental process was applied.

- D: The result is publishable in its own right as a new scientific result—independent of the fact that the result was mechanically created.

All of the results in this chapter, with the exception of the result for the scaling Majority-ON problem, meet criterion B. The results for the OR, AND/OR, and gate communication problems also meet criterion D, as established by publications in physics venues (Barnum et al., 2000, Spector and Bernstein, 2003).

The solution to the 1-bit Deutsch-Jozsa (XOR) problem appears simple in retrospect, but one must remember that this surprising and powerful effect went unnoticed for the first 60 years following the development of the underlying quantum mechanics. And even now it is counterintuitive to most people. It is true that much of the intelligence behind this result lies in the human discovery that the problem was worth posing in the first place, but the steps from the problem statement to a solution are nonetheless non-trivial. The fact that genetic programming can proceed automatically to a solution when provided only with the problem statement and a generic set of quantum gates is therefore significant.

Similar comments apply to the result for Grover's database search problem. Although a human being (Lov Grover) was responsible for the insight that quantum computers could outperform classical computers on this problem, the production of a better-than-classical quantum algorithm for the problem is nonetheless difficult and represents a significant achievement for an automatic programming system. It is also noteworthy that the first time this result was produced by genetic programming it exceeded the expectations of the person performing the experiment (the author of this book), who had naively assumed that the $\sqrt{n}$ improvement would allow only for a two-oracle-call solution. Although the zero-error, single-call solution added nothing to the state of the art in quantum computing, its possibility was news to the designer and user of the automatic quantum computer programming system (who was at that time new to the field of quantum computing). This is important because it demonstrates that the system can produce knowledge beyond that possessed by the system designers or users.

The results on the OR and AND/OR problems were published in *Journal of Physics A: Mathematical and General* on the strength of their contributions to the theory of quantum computing, not on the basis of their production by mechanical means. Although the article does briefly describe the genetic programming methodology that produced the results, neither the article's title nor its abstract mention how the results were produced. The novel methodology by which these results were produced would probably not, by itself, warrant publication in this particular journal, which routinely publishes articles on quantum complexity theory but not on the design of automatic programming systems. The fact that these results were published in a high-quality, peer-reviewed *physics* journal demonstrates that the approach to automatic quantum computer programming described in this book can produce new scientific results that are on par with those produced by human scientists.

The result on the Scaling Majority-ON problem is of more limited significance; it serves only to demonstrate how genetic programming can be employed to find scalable solutions to problems that have instances of various sizes. But the result itself is not better than classical, and it is also fairly obvious. It is significant only insofar as it points the way to more ambitious applications of genetic programming to other problems in the future.

Several of the results on classical communication via particular quantum gates are new scientific contributions, significant independent of the means by which they were produced. Evidence for this is their publication in the *Proceedings of the Sixth International Conference on Quantum Communication, Measurement, and Computing.* It is also noteworthy that in this case the genetic programming system was employed in a role similar to that of a scientific colleague. The system was used first to investigate a particular question ("Can classical information be transmitted via a SMOLIN gate?") but its result ("Yes") was not the end of the story; the details of the result inspired a round of human analysis and the production of new questions for the system. Results of the runs on these secondary questions have led to further analysis and insights. This work is ongoing and additional publications in the physics literature are expected in the future (Bennett et al., 2004).

Chapter 9

CONCLUSIONS AND PROSPECTS

Quantum computing is an exciting frontier of computer science that may, if the aspirations of its proponents are fully realized, provide humanity with truly awesome computational power. At present, however, we have only hints of the power that may be available, and we have only begun to grapple with the practical problems involved in the construction of large-scale quantum computers.

Many open problems in quantum computing and quantum information theory can be formulated as searches for quantum programs that have particular properties. In other words, they can be thought of as programming problems, and more specifically as quantum computer programming problems. Unfortunately, quantum computers are counterintuitive and difficult to program. But fortunately we can adapt existing automatic programming technologies to help us to search for quantum programs. A successful automatic quantum computer programming system could contribute to our understanding of quantum computing in several ways.

Genetic and evolutionary computation technologies — in particular genetic programming technologies — provide powerful methods for automatic programming. Recent advances in genetic programming techniques enable the evolution of complex programs that solve difficult, real-world problems. When augmented with quantum computer simulation facilities these systems can be used for automatic quantum computer programming, thereby aiding the exploration of quantum computing.

This book described several ways in which genetic programming can support automatic quantum computer programming, culminating in a set of specific techniques and examples. These examples, along with related results produced by other researchers, (for example, Williams and Gray, 1999; Surkan and Khuskivadze, 2001; Leier and Banzhaf, 2003a; Leier and Banzhaf, 2003b; Perkowski et al., 2003; Massey et al., 2004), provide reasons to expect more dramatic discoveries from these or related techniques in the future.

Straightforward improvements to many of the technologies presented here should extend their capabilities in significant ways. The specific systems described here, including QGAME and PushGP, were developed as part of an exploratory research process and were not optimized for execution speed. Most of them were written in un-optimized Common Lisp for the sake of rapid prototyping and experimentation, but faster versions of PushGP and QGAME, written in C++, have just become available. These improvements, coupled with deployment across larger networked clusters of computers (as described in Chapter 4), should significantly increase the size of the quantum systems that can be simulated and the reach of the genetic programming searches that can be conducted. Improvements to the underlying quantum computer simulation algorithms, for example those that avoid exponential slowdowns for classical segments of quantum programs, may allow for further scale-ups.

An important question *not* addressed in this book is "How can we determine what open problems in quantum computing are best addressed by means of genetic programming?" This is a difficult question to answer without deep knowledge both in quantum computing and in genetic programming. It is hoped that this book will encourage more people to seek such knowledge in both areas, and subsequently to apply genetic programming to new problems in quantum computing.

The "low-hanging fruit" for future applications are clearly other "small n" problems. Any open problem that can be resolved, one way or the other, with the discovery of a single, small quantum program is worth considering as a candidate for solution via genetic programming. The techniques for evolving such single-size programs are straightforward and the exponential overhead for quantum computer simulation is manageable for systems with small numbers of qubits.

More interesting, and more challenging, are problems that involve programs that must be scaled up for various values of n. There are many such problems — several of the important open questions in quantum computing concern the asymptotic computational complexity of problems as they grow in size. A basic technique for approaching these problems was presented here, but the exponential overhead for quantum computer simulation may limit the use of this technique. Other advances may be necessary to achieve significant scaling results.

For the longer term, it is interesting to speculate about new sorts of applications that might become practical using variants of the quantum efficiencies that have already been discovered, and to consider the ways in which automatic quantum computer programming technologies might help us to design such applications.

Grover's search algorithm has many obvious applications, to which it can provide a quadratic speedup. It is also possible that Grover's ideas can be extended to provide more substantial speedups for certain specialized searches.[1] A great deal of work in artificial intelligence views *all* interesting computation as forms of search, and these ideas might be used, in conjunction with refined quantum search algorithms, to support an array of efficient quantum artificial intelligence technologies. Indeed, as mentioned above one can even view automatic programming as a form of search, and the notion of using quantum computers to speed up automatic programming technologies such as genetic programming has been raised in the literature several times (Spector et al., 1998; Spector et al., 1999b; Rylander et al., 2001). One specialized form of search that has wide application in AI is search over AND/OR trees, which also form the foundation of some kinds of logic programming; one might therefore speculate that the quantum speedups discovered for the AND/OR problem may support some form of "quantum logic machine."

Other obvious areas for applications include numerical analysis and cryptography, where we may expect techniques related to Shor's quantum Fourier transform and factoring algorithms to find new uses. We might further speculate that technologies rooted in massive parallelism, such as neural networks, will benefit rather directly from the form of exponential parallelism provided by quantum computers. By similar logic we might expect technologies rooted in the manipulation of probabilities, such as Bayesian networks, to benefit from the unique probability-processing features of quantum computers. The capability of quantum computers to represent superpositions of multiple states may also have unexpected applications; for example, a recent Ph.D. dissertation claims that quantum mechanical superpositions may have an important role to play in natural language processing (Chen, 2002).

Each of these speculations leads in turn to a new set of questions, and it is possible that many of these questions will be answered in the future by automatic quantum computer programming technologies.

---

[1] For some initial steps in this direction see (Hogg, 1998; Hogg, 2000).

Appendix A
QGAME source code

This appendix contains Common Lisp source code for the core elements of the QGAME quantum computer simulator. It omits much of the program documentation and also some of the system's advanced features (such as the algorithms for gate compression). Fullly documented source code for this and other versions of QGAME can be obtained online.[1]

```
;;;;;;;;;;;;;;;;;;;;;;;;;;;;;;;;;;;;;;;;;;;;;;;;;;;;;;;;;;;;;;;;;;;;;;;;;
;; qgame.lisp
;; A minimal and lightly documented version of QGAME, the Quantum Gate
;; And Measurement Emulator, implemented in Common Lisp and prepared for
;; inclusion in:
;;
;; AUTOMATIC QUANTUM COMPUTER PROGRAMMING: A GENETIC PROGRAMMING
;; APPROACH, by Lee Spector, published by Kluwer Academic Publishers
;;
;; Full source and documentation is available from:
;;
;; http://hampshire.edu/lspector/qgame.html
;;
;; c) 1999-2004, Lee Spector, lspector@hampshire.edu
```

[1] http://hampshire.edu/lspector/qgame.html

```
;;;;;;;;;;;;;;;;;;;;;;;;;;;;;;;;;;;;;;;;;;;;;;;;;;;;;;;;;;;;;;;;;;;;;;;;;
;; class definition for a quantum system

(defclass quantum-system ()
  (;; the number of qubits in the system
   (number-of-qubits :accessor number-of-qubits
                     :initarg :number-of-qubits)
   ;; an array of amplitudes
   (amplitudes :accessor amplitudes
               :initarg :amplitudes
               :initform nil)
   ;; the probability for having reached this system
   ;; in the first place
   (prior-probability :accessor prior-probability
                      :initarg :prior-probability
                      :initform 1)
   ;; the number of oracle calls that have been made
   ;; in the history of this system
   (oracle-count :accessor oracle-count
                 :initarg :oracle-count
                 :initform 0)
   ;; a list of measurements and their results in
   ;; the history of this system
   (measurement-history :accessor measurement-history
                        :initarg :measurement-history
                        :initform nil)
   ;; a list of all instructions executed in the
   ;; history of this system
   (instruction-history :accessor instruction-history
                        :initarg :instruction-history
                        :initform nil)
   ;; the program yet to be executed by this system
   ;; (if it hasn't yet terminated)
   (program :accessor program
            :initarg :program
            :initform nil)
   ;; the following are just for convenience
   ;; a list of all valid qubit indices
   (qubit-numbers :accessor qubit-numbers)
   ;; address storage, used for looping through qubits
   (amplitude-address :accessor amplitude-address)))
```

```
(defmethod initialize-instance
           :after ((qsys quantum-system) &rest args)
  "An initializer for quantum systems."
  (declare (ignore args))
  (let ((num-qubits (number-of-qubits qsys)))
    ;; if there are no amplitudes yet then initialize to |00...0>
    (unless (amplitudes qsys)
      (setf (amplitudes qsys)
            (let ((amps (make-array (expt 2 num-qubits)
                                    :initial-element 0.0L0)))
              (setf (aref amps 0) 1.0L0) ;; start in zero state
              amps)))
    ;; initilize list of valid qubit indices
    (setf (qubit-numbers qsys)
          (let ((all nil))
            (dotimes (i num-qubits) (push i all))
            (reverse all)))
    ;; initialize address register for amplitudes
    (setf (amplitude-address qsys)
          (make-array num-qubits :initial-element 0))))

;;;;;;;;;;;;;;;;;;;;;;;;;;;;;;;;;;;;;;;;;;;;;;;;;;;;;;;;;;;;;;;;;;;;;;;;;;;;
;; quantum computer manipulation utilities

(defun set-address-components (qsys count qubits)
  "Sets (amplitude-address qsys) to refer to a particular amplitude, as
indicated by the bits in the integer count."
  (dotimes (i (length qubits))
    (setf (aref (amplitude-address qsys) (nth i qubits))
          (if (logbitp i count) 1 0))))

(defun map-qubit-combinations (qsys function qubits)
  "Calls function once for each of the 1/0 combinations of the provided
qubits, with the right-most qubit varying the fastest."
  (setq qubits (reverse qubits))
  (let ((number-of-iterations (expt 2 (length qubits))))
    (dotimes (i number-of-iterations)
      (set-address-components qsys i qubits)
      (funcall function))))

(defun get-addressed-amplitude (qsys)
  "Returns the amplitude that is currently addressed
by (amplitude-address qsys)"
  (let ((numerical-address 0))
    (dotimes (i (number-of-qubits qsys))
      (unless (zerop (aref (amplitude-address qsys) i))
        (incf numerical-address (expt 2 i))))
    (aref (amplitudes qsys) numerical-address)))
```

```
(defun set-addressed-amplitude (qsys new-value)
  "Sets the amplitude currently addressed by (amplitude-address qsys)
to new-value."
  (let ((numerical-address 0))
    (dotimes (i (number-of-qubits qsys))
      (unless (zerop (aref (amplitude-address qsys) i))
        (incf numerical-address (expt 2 i))))
    (setf (aref (amplitudes qsys) numerical-address) new-value)))

(defun matrix-multiply (matrix column)
  "Multiplies the given square matrix by the given column (assumed
to be the right length) and returns the resulting column."
  (let ((matrix-size (car (array-dimensions matrix)))
        (result nil))
    (dotimes (i matrix-size)
      (push (let ((element 0))
              (dotimes (j matrix-size)
                (incf element (* (aref matrix i j) (nth j column))))
              element)
            result))
    (reverse result)))

(defun extract-column (qsys qubits-to-vary)
  "Returns a column from the amplitudes obtained by varying the listed
qubits, with the right-most qubit varying the fastest."
  (let ((col nil))
    (map-qubit-combinations
     qsys
     #'(lambda ()
         (push (get-addressed-amplitude qsys) col))
     qubits-to-vary)
    (reverse col)))

(defun install-column (qsys column qubits-to-vary)
  "Installs the given column in the amplitude positions obtained by
varying the listed qubits, with the right-most qubit varying the
fastest."
  (map-qubit-combinations
   qsys
   #'(lambda ()
       (set-addressed-amplitude qsys (car column))
       (setq column (cdr column)))
   qubits-to-vary))
```

```
(defun apply-operator (qsys operator qubits)
  "Applies the given matrix-form operator to the given qubits."
  (map-qubit-combinations
   qsys
   #'(lambda ()
       (let* ((pre-column (extract-column qsys qubits))
              (post-column (matrix-multiply operator pre-column)))
         (install-column qsys post-column qubits)))
   (set-difference (qubit-numbers qsys) qubits))
  qsys)

(defun qc-output-probabilities (qsys qubits)
  "Returns a list of the probabilities for all combinations for the
given qubits, in binary order with the right-most qubit varying fastest."
  (let ((probabilities nil)
        (other-qubits (set-difference (qubit-numbers qsys) qubits)))
    (map-qubit-combinations
     qsys
     #'(lambda ()
         (push (let ((probability 0))
                 (map-qubit-combinations
                  qsys
                  #'(lambda ()
                      (incf probability
                            (expt (abs (get-addressed-amplitude qsys))
                                  2)))
                  other-qubits)
                 probability)
               probabilities))
     qubits)
    (reverse probabilities)))

(defun multi-qsys-output-probabilities (qsys-list qubits)
  "Returns a list of the probabilities for all combinations for the
given qubits, in binary order with the right-most qubit varying fastest.
This function takes a LIST of quantum systems as input and sums the
results across all systems."
  (let ((probabilities
         (mapcar #'(lambda (qsys)
                     (qc-output-probabilities qsys qubits))
                 qsys-list)))
    (labels ((add-lists (l1 l2)
               (if (null l1)
                 nil
                 (cons (+ (first l1) (first l2))
                       (add-lists (rest l1) (rest l2))))))
      (reduce #'add-lists probabilities))))
```

```
(defun expected-oracles (qsys-list)
  "Returns the expected number of oracle calls for the given
set of quantum systems."
  (reduce #'+
         (mapcar #'(lambda (qsys)
                     (* (prior-probability qsys)
                        (oracle-count qsys)))
                 qsys-list)))

;;;;;;;;;;;;;;;;;;;;;;;;;;;;;;;;;;;;;;;;;;;;;;;;;;;;;;;;;;;;;;;;;;;;;;;;;
;; oracle gates

(defun binary-operator-matrix (tt-right-column)
  "Returns a matrix operator for a binary function with the
given tt-right-column as the right column of its truth table."
  (let* ((column-length (length tt-right-column))
         (operator-size (* 2 column-length))
         (matrix (make-array (list operator-size operator-size)
                             :initial-element 0)))
    (dotimes (i column-length)
      (let ((offset (* i 2)))
        (if (zerop (nth i tt-right-column))
          (setf (aref matrix offset offset) 1
                (aref matrix (1+ offset) (1+ offset)) 1)
          (setf (aref matrix offset (1+ offset)) 1
                (aref matrix (1+ offset) offset) 1))))
    matrix))

(defun oracle (qsys tt-right-column &rest qubits)
  "Applies the oracle operator built from tt-right-column, which
is the right column of the corresponding truth table."
  (incf (oracle-count qsys))
  (apply-operator
   qsys
   (binary-operator-matrix tt-right-column)
   qubits))

(defun limited-oracle (qsys max-calls tt-right-column &rest qubits)
  "If (oracle-count qsys) is less than max-calls then this applies
the oracle operator built from tt-right-column, which is the right
column of the corresponding truth table. Otherwise this does nothing."
  (if (< (oracle-count qsys) max-calls)
    (progn (incf (oracle-count qsys))
           (apply-operator
            qsys
            (binary-operator-matrix tt-right-column)
            qubits))
    qsys))
```

```
;;;;;;;;;;;;;;;;;;;;;;;;;;;;;;;;;;;;;;;;;;;;;;;;;;;;;;;;;;;;;;;;;;;;;;;;;;;;;;;
;; other quantum gates

(defun qnot (qsys q)
  "Quantum NOT gate"
  (apply-operator qsys
                  #2A((0 1)
                      (1 0))
                  (list q)))

(defun cnot (qsys q1 q2)
  "Quantum Controlled NOT gate"
  (apply-operator qsys
                  #2A((1 0 0 0)
                      (0 1 0 0)
                      (0 0 0 1)
                      (0 0 1 0))
                  (list q1 q2)))

(defun srn (qsys q)
  "Quantum Square-Root-of-NOT gate"
  (apply-operator
   qsys
   (make-array
    '(2 2)
    :initial-contents
    (list (list (/ 1 (sqrt 2.0L0))  (- (/ 1 (sqrt 2.0L0))))
          (list (/ 1 (sqrt 2.0L0))  (/ 1 (sqrt 2.0L0)))))
   (list q)))

(defun hadamard (qsys q)
  "Quantum Hadamard gate"
  (apply-operator
   qsys
   (make-array
    '(2 2)
    :initial-contents
    (list (list (/ 1 (sqrt 2.0L0))  (/ 1 (sqrt 2.0L0)))
          (list (/ 1 (sqrt 2.0L0))  (- (/ 1 (sqrt 2.0L0))))))
   (list q)))
```

```
(defun u-theta (qsys q theta)
  "Quantum U-theta (rotation) gate"
  (apply-operator
   qsys
   (make-array
    '(2 2)
    :initial-contents
    (list (list (cos theta)  (sin theta))
          (list (- (sin theta))  (cos theta))))
   (list q)))

(defun cphase (qsys q1 q2 alpha)
  "Quantum controlled phase gate"
  (apply-operator
   qsys
   (make-array
    '(4 4)
    :initial-contents
    (list (list 1 0 0 0)
          (list 0 1 0 0)
          (list 0 0 1 0)
          (list 0 0 0 (exp (* (sqrt -1.0L0) alpha)))))
   (list q1 q2)))

;; U2 =  U(phi) * R(theta) * U(psi) * exp(i alpha)I
;; where  U(a) = e^(-ia) 0
;;              0       e^(ia)
;; and    R(a) = cos(a) sin(-a)
;;              sin(a) cos(a)
;; This is all pre-multiplied in the following code

(defun u2 (qsys q phi theta psi alpha)
  "Quantum U2 gate, implemented as:
e^(i(-phi-psi+alpha))*cos(theta)  e^(i(-phi+psi+alpha))*sin(-theta)
e^(i(phi-psi+alpha))*sin(theta)   e^(i(phi+psi+alpha))*cos(theta)"
  (apply-operator
   qsys
   (let ((i (sqrt -1.0L0)))
     (make-array
      '(2 2)
      :initial-contents
      (list (list (* (exp (* i (+ (- phi) (- psi) alpha))) (cos theta))
                  (* (exp (* i (+ (- phi) psi alpha))) (sin (- theta))))
            (list (* (exp (* i (+ phi (- psi) alpha))) (sin theta))
                  (* (exp (* i (+ phi psi alpha))) (cos theta)))
            )))
   (list q)))
```

```
(defun swap (qsys q1 q2)
  "A quantum gate that swaps the amplitudes for the two specified
qubits."
  (apply-operator
   qsys
   (make-array
    '(4 4)
    :initial-contents
    (list (list 1 0 0 0)
          (list 0 0 1 0)
          (list 0 1 0 0)
          (list 0 0 0 1)))
   (list q1 q2)))

;;;;;;;;;;;;;;;;;;;;;;;;;;;;;;;;;;;;;;;;;;;;;;;;;;;;;;;;;;;;;;;;;;;;;;;;;;;;;;
;; utilities for measurement and branching

(defun end (qsys)
  "Marks the end of a measurement branch; has no effect when used
in a quantum program in any other context."
  qsys)

(defun distance-to-next-unmatched-end
       (list &optional
             (num-measures 0) (num-ends 0) (distance-so-far 0))
  "Returns 0 if there is no unmatched (end) in list;
otherwise  returns the number of instructions to the next
unmatched (end) (counting the (end))."
  (if (null list)
    0
    (if (eq (caar list) 'end)
      (if (zerop num-measures)
        (+ 1 distance-so-far)
        (if (oddp num-ends) ;; then this one closes a measure
          (distance-to-next-unmatched-end (cdr list)
                                          (- num-measures 1)
                                          (- num-ends 1)
                                          (+ 1 distance-so-far))
          (distance-to-next-unmatched-end (cdr list)
                                          num-measures
                                          (+ num-ends 1)
                                          (+ 1 distance-so-far))))
      (if (eq (caar list) 'measure)
        (distance-to-next-unmatched-end (cdr list)
                                        (+ num-measures 1) num-ends
                                        (+ 1 distance-so-far))
        (distance-to-next-unmatched-end (cdr list)
                                        num-measures num-ends
                                        (+ 1 distance-so-far))))))
```

```
(defun without-if-branch (program)
  "Assuming that a MEASURE form has just been removed from the
given program, returns the remainder of the program without the
IF (measure-1) branch."
  (let* ((distance-to-first-unmatched-end
          (distance-to-next-unmatched-end program))
         (distance-from-first-to-second-unmatched-end
          (distance-to-next-unmatched-end
           (nthcdr distance-to-first-unmatched-end program))))
    (if (zerop distance-to-first-unmatched-end)
      ;; it's all the if part
      nil
      ;; there is some else part
      (if (zerop distance-from-first-to-second-unmatched-end)
        ;; the else never ends
        (subseq program distance-to-first-unmatched-end)
        ;; the else does end
        (append
         (subseq program
                 distance-to-first-unmatched-end
                 (+ distance-to-first-unmatched-end
                    distance-from-first-to-second-unmatched-end
                    -1))
         (subseq program (+ distance-to-first-unmatched-end
                            distance-from-first-to-second-unmatched-end
                            )))))))

(defun without-else-branch (program)
  "Assuming that a MEASURE form has just been removed from the
given program, returns the remainder of the program without the
ELSE (measure-0) branch."
  (let* ((distance-to-first-unmatched-end
          (distance-to-next-unmatched-end program))
         (distance-from-first-to-second-unmatched-end
          (distance-to-next-unmatched-end
           (nthcdr distance-to-first-unmatched-end program))))
    (if (zerop distance-to-first-unmatched-end)
      ;; it's all the if part
      program
      ;; there is some else part
      (if (zerop distance-from-first-to-second-unmatched-end)
        ;; the else never ends
        (subseq program 0 (- distance-to-first-unmatched-end 1))
        ;; the else does end
        (append
         (subseq program 0 (- distance-to-first-unmatched-end 1))
         (subseq program (+ distance-to-first-unmatched-end
                            distance-from-first-to-second-unmatched-end
                            )))))))
```

```
(defun force-to (measured-value qubit qsys)
  "Collapses a quantum system to the provided measured-value for the
provided qubit."
  (map-qubit-combinations
   qsys
   #'(lambda ()
       (let* ((pre-column (extract-column qsys (list qubit)))
              (new-column (case measured-value
                            (0 (list (first pre-column) 0))
                            (1 (list 0 (second pre-column))))))
         (install-column qsys new-column (list qubit))))
   (remove qubit (qubit-numbers qsys)))
  qsys)


;;;;;;;;;;;;;;;;;;;;;;;;;;;;;;;;;;;;;;;;;;;;;;;;;;;;;;;;;;;;;;;;;;;;;;;;;;;;;;;
;; top level functions


(defun execute-quantum-program (pgm num-qubits
                                    &optional (oracle-tt nil))
  "Executes the provide quantum program with the specified number of
qubits and the provided oracle truth table, returning a list of the
resulting quantum systems."
  (run-qsys (make-instance 'quantum-system
              :number-of-qubits num-qubits
              :program (subst oracle-tt 'ORACLE-TT pgm))))
```

```
(defun run-qsys (qsys)
  "Takes a quantum system and returns the list of quantum systems that
results from the execution of its program."
  (if (or (null (program qsys))
          (zerop (prior-probability qsys)))
    (list qsys)
    (let ((instruction (first (program qsys))))
      (setf (instruction-history qsys)
            (append (instruction-history qsys) (list instruction)))
      (if (eq (first instruction) 'halt)
        (list qsys)
        (if (eq (first instruction) 'measure)
          (let* ((measurement-qubit (second instruction))
                 (probabilities (qc-output-probabilities
                                  qsys (list measurement-qubit))))
            (append (run-qsys ;; 1 branch
                     (force-to
                      1 measurement-qubit
                      (make-instance 'quantum-system
                        :number-of-qubits (number-of-qubits qsys)
                        :amplitudes (copy-seq (amplitudes qsys))
                        :prior-probability (second probabilities)
                        :oracle-count (oracle-count qsys)
                        :measurement-history
                        (append (measurement-history qsys)
                                (list (list measurement-qubit 'is 1)))
                        :instruction-history (instruction-history qsys)
                        :program (without-else-branch
                                   (rest (program qsys))))))
                    (run-qsys ;; 0 branch
                     (force-to
                      0 measurement-qubit
                      (make-instance 'quantum-system
                        :number-of-qubits (number-of-qubits qsys)
                        :amplitudes (copy-seq (amplitudes qsys))
                        :prior-probability (first probabilities)
                        :oracle-count (oracle-count qsys)
                        :measurement-history
                        (append (measurement-history qsys)
                                (list (list measurement-qubit 'is 0)))
                        :instruction-history (instruction-history qsys)
                        :program (without-if-branch
                                   (rest (program qsys))))))))
          (let ((resulting-sys (apply (first instruction)
                                      (cons qsys (rest instruction)))))
            (setf (program resulting-sys)(rest (program resulting-sys)))
            (run-qsys resulting-sys)))))))
```

```
(defun test-quantum-program
       (pgm &key num-qubits cases final-measurement-qubits threshold
            (inspect nil) (debug 0))
  "The top-level function to evaluate a quantum program relative to
a list of (oracle value) cases. Returns a list of: misses, max-error,
average-error, max-expected-oracles, and average-expected-oracles.
See complete documentation for a more complete explanation of the
arguments and return values."
  (let ((misses 0)
        (max-error 0)
        (total-error 0)
        (average-error 0)
        (max-expected-oracles 0)
        (total-expected-oracles 0)
        (average-expected-oracles 0)
        (num-cases (length cases)))
    (dolist (case cases)
      (let* ((resulting-systems
               (execute-quantum-program pgm num-qubits (first case)))
             (raw-error
               (- 1.0
                  (nth (second case)
                       (multi-qsys-output-probabilities
                         resulting-systems
                         final-measurement-qubits))))
             (expected-oracles (expected-oracles resulting-systems)))
        (if (> raw-error threshold) (incf misses))
        (incf total-error raw-error)
        (when (> raw-error max-error)
          (setq max-error raw-error))
        (incf total-expected-oracles expected-oracles)
        (when (> expected-oracles max-expected-oracles)
          (setq max-expected-oracles expected-oracles))
        (when (>= debug 2)
          (format t "~%---~%Case:~A, Error:~,5F" case raw-error))
        (when inspect (inspect resulting-systems))))
    (setq average-error (/ total-error num-cases))
    (setq average-expected-oracles (/ total-expected-oracles num-cases))
    (when (>= debug 1)
      (format t "~%~%Misses:~A" misses)
      (format t "~%Max error:~A" max-error)
      (format t "~%Average error:~A" (float average-error))
      (format t "~%Max expected oracles:~A" max-expected-oracles)
      (format t "~%Average expected oracles:~A"
              (float average-expected-oracles)))
    (list misses max-error average-error max-expected-oracles
          average-expected-oracles)))
```

References

Albert, D. Z. (1992). *Quantum Mechanics and Experience.* Harvard University Press, Cambridge, Massachusetts.

Angeline, P. J. and Kinnear, Jr., K. E., editors (1996). *Advances in Genetic Programming 2.* MIT Press, Cambridge, MA, USA.

Angeline, P. J. and Pollack, J. B. (1992). The evolutionary induction of subroutines. In *Proceedings of the Fourteenth Annual Conference of the Cognitive Science Society,* Bloomington, Indiana, USA. Lawrence Erlbaum.

Angeline, P. J. and Pollack, J. B. (1993). Evolutionary module acquisition. In Fogel, D. and Atmar, W., editors, *Proceedings of the Second Annual Conference on Evolutionary Programming,* pages 154–163, La Jolla, CA, USA.

Banzhaf, W., Nordin, P., Keller, R. E., and Francone, F. D. (1998). *Genetic Programming – An Introduction; On the Automatic Evolution of Computer Programs and its Applications.* Morgan Kaufmann, dpunkt.verlag.

Barenco, A., Bennett, C. H., Cleve, R., DiVincenzo, D. P., Margolus, N., Shor, P., Sleator, T., Smolin, J. A., and Weinfurter, H. (1995). Elementary gates for quantum computation. *Physical Review A,* 52:3457–3467.

Barnum, H., Bernstein, H. J., and Spector, L. (2000). Quantum circuits for OR and AND of ORs. *Journal of Physics A: Mathematical and General,* 33(45):8047–8057.

Beckman, D., Chari, A. N., Devabhaktuni, S., and Preskill, J. (1996). Efficient networks for quantum factoring. Technical Report CALT-68-2021, California Institute of Technology. `http://xxx.lanl.gov/abs/quant-ph/9602016`.

Bell, J. S. (1993). *Speakable and unspeakable in quantum mechanics.* Cambridge University Press, Cambridge.

Bennett, C. H. (1999). Quantum information theory. In Hey, A. J. G., editor, *Feynman and Computation: Exploring the Limits of Computers,* pages 177–190. Persus Books, Reading, Massachusetts.

Bennett, C. H., Bernstein, H. J., Harrow, A., Leung, D. W., Smolin, J. A., and Spector, L. (2004). Evidence for unequal efficiencies of some quantum gates for forward communication, backward communication and entanglement generation, discovered in part by genetic programming. (in preparation).

Brooks, M., editor (1999). *Quantum Computing and Communications.* Springer-Verlag, London.

Brown, J. (2000). *Minds, Machines and the Multiverse: The Quest for the Quantum Computer.* Simon & Schuster.

Burke, E., Gustafson, S., and Kendall, G. (2002a). A survey and analysis of diversity measures in genetic programming. In Langdon, W. B., Cantú-Paz, E., Mathias, K., Roy, R., Davis, D., Poli, R., Balakrishnan, K., Honavar, V., Rudolph, G., Wegener, J., Bull, L., Potter, M. A., Schultz, A. C., Miller, J. F., Burke, E., and Jonoska, N., editors, *GECCO 2002: Proceedings of the Genetic and Evolutionary Computation Conference*, pages 716–723, New York. Morgan Kaufmann Publishers.

Burke, E., Gustafson, S., Kendall, G., and Krasnogor, N. (2002b). Advanced population diversity measures in genetic programming. In Guervós, J.-J. M., Adamidis, P., Beyer, H.-G., nas, J.-L. F.-V., and Schwefel, H.-P., editors, *Parallel Problem Solving from Nature - PPSN VII*, number 2439 in Lecture Notes in Computer Science, LNCS, page 341 ff., Granada, Spain. Springer-Verlag.

Chen, J. C. (2002). *Quantum Computation and Natural Language Processing*. PhD thesis, Department of Computer Science, University of Hamburg, Vogt-Kölln-Strae 30, D-22527 Hamburg, Germany.
`http://nats-www.informatik.uni-hamburg.de/~joseph/dis/`.

Christensen, S. and Oppacher, F. (2001). What can we learn from no free lunch? A first attempt to characterize the concept of a searchable function. In Spector, L., Goodman, E. D., Wu, A., Langdon, W. B., Voigt, H.-M., Gen, M., Sen, S., Dorigo, M., Pezeshk, S., Garzon, M. H., and Burke, E., editors, *Proceedings of the Genetic and Evolutionary Computation Conference (GECCO-2001)*, pages 1219–1226, San Francisco, California, USA. Morgan Kaufmann.

Cramer, N. L. (1985). A representation for the adaptive generation of simple sequential programs. In Grefenstette, J. J., editor, *Proceedings of an International Conference on Genetic Algorithms and the Applications*, pages 183–187, Carnegie-Mellon University, Pittsburgh, PA, USA.

Crawford-Marks, R. and Spector, L. (2002). Size control via size fair genetic operators in the PushGP genetic programming system. In Langdon, W. B., Cantú-Paz, E., Mathias, K., Roy, R., Davis, D., Poli, R., Balakrishnan, K., Honavar, V., Rudolph, G., Wegener, J., Bull, L., Potter, M. A., Schultz, A. C., Miller, J. F., Burke, E., and Jonoska, N., editors, *GECCO 2002: Proceedings of the Genetic and Evolutionary Computation Conference*, pages 733–739, New York. Morgan Kaufmann Publishers.

Deutsch, D. (1997). *The Fabric of Reality*. Penguin Books.

Deutsch, D. and Jozsa, R. (1992). Rapid solution of problems by quantum computation. *Proceedings of the Royal Society of London Ser.A*, A439:553–558.

Droste, S., Jansen, T., and Wegener, I. (1999). Perhaps not a free lunch but at least a free appetizer. In Banzhaf, W., Daida, J., Eiben, A. E., Garzon, M. H., Honavar, V., Jakiela, M., and Smith, R. E., editors, *Proceedings of the Genetic and Evolutionary Computation Conference*, volume 1, pages 833–839, Orlando, Florida, USA. Morgan Kaufmann.

Edmonds, B. (2001). Meta-genetic programming: Co-evolving the operators of variation. *Elektrik*, 9(1):13–29. Turkish Journal Electrical Engineering and Computer Sciences.

Ekart, A. and Nemeth, S. Z. (2001). Selection based on the pareto nondomination criterion for controlling code growth in genetic programming. *Genetic Programming and Evolvable Machines*, 2(1):61–73.

Elitzur, A. C. and Vaidman, L. (1993). Quantum mechanical interaction-free measurements. *Foundation of Physics*, 23:987–997.
`http://arxiv.org/abs/hep-th/9305002`.

Feynman, R. P. (1985). *QED: The Strange Theory of Light and Matter.* Princeton University Press.

Feynman, R. P. (1996). *Feynman Lectures on Computation.* Perseus Publishing, Cambridge, Massachusetts.

Fogel, D. B. and Atmar, J. W. (1990). Comparing genetic operators with Gaussian mutations in simulated evolutionary processes using linear systems. *Biological Cybernetics*, 63(2):111–114.

Goldberg, D. E. (1989). *Genetic Algorithms in Search, Optimization, and Machine Learning.* Addison-Wesley.

Graham, P. (1994). *On Lisp: advanced techniques for Common Lisp.* Prentice-Hall, Englewood Cliffs, NJ 07632, USA.

Grover, L. K. (1997). Quantum mechanics helps in searching for a needle in a haystack. *Physical Review Letters*, pages 325–328.

Gruau, F. (1994). Genetic micro programming of neural networks. In Kinnear, Jr., K. E., editor, *Advances in Genetic Programming*, chapter 24, pages 495–518. MIT Press.

Gruska, J. (1999). *Quantum Computing.* McGraw-Hill Publishing Company, Maidenhead, Berkshire.

Hallgren, S. (2002). Polynomial-time quantum algorithms for pell's equation and the principal ideal problem. In *Proceedings of the 34th ACM Symposium on Theory of Computing.*

Hallgren, S., Russell, A., and Ta-Shma, A. (2003). The hidden subgroup problem and quantum computation using group representations. *SIAM J. Comput.*, 32(4):916–934.

Hey, A. J. G., editor (1999). *Feynman and Computation: Exploring the Limits of Computers.* Persus Books, Reading, Massachusetts.

Hogg, T. (1998). Highly structured searches with quantum computers. *Physical Review Letters*, 80:2473–2476.

Hogg, T. (2000). Quantum search heuristics. *Physical Review A*, 61:052311.

Holland, J. H. (1992). *Adaptation in Natural and Artificial Systems: An introductory analysis with applications to biology, control, and artificial intelligence.* The MIT Press, Cambridge, Massachusetts. First edition ©1975.

Igel and Toussaint (2003). On classes of functions for which no free lunch results hold. *IPL: Information Processing Letters*, 86.

Jozsa, R. (1997). Entanglement and quantum information. In Hugett, S., Mason, L., Todd, K. P., Tsou, S. T., and Woodhouse, N. J., editors, *Geometric Issues in the Foundations of Science.* Oxford University Press. `http://arXiv.org/quant-ph/9707034`.

Kinnear, Jr., K. E., editor (1994a). *Advances in Genetic Programming.* MIT Press, Cambridge, MA.

Kinnear, Jr., K. E. (1994b). Alternatives in automatic function definition: A comparison of performance. In Kinnear, Jr., K. E., editor, *Advances in Genetic Programming*, chapter 6, pages 119–141. MIT Press.

Klein, J. (2002). BREVE: a 3d environment for the simulation of decentralized systems and artificial life. In Standish, R. K., Bedau, M. A., and Abbass, H. A., editors, *Proceedings of Artificial Life VIII, the 8th International Conference on the Simulation and Synthesis of Living Systems*, pages 329–334. The MIT Press. `http://www.spiderland.org/breve/breve-klein-alife2002.pdf`.

Koza, J. (1990). Genetic programming: A paradigm for genetically breeding populations of computer programs to solve problems. Technical Report STAN-CS-90-1314, Dept. of Computer Science, Stanford University.

Koza, J. R. (1992). *Genetic Programming: On the Programming of Computers by Means of Natural Selection.* MIT Press, Cambridge, MA, USA.

Koza, J. R. (1994). *Genetic Programming II: Automatic Discovery of Reusable Programs.* MIT Press, Cambridge Massachusetts.

Koza, J. R., David Andre, Bennett III, F. H., and Keane, M. (1999). *Genetic Programming III: Darwinian Invention and Problem Solving.* Morgan Kaufman.

Koza, J. R., Keane, M. A., Streeter, M. J., Mydlowec, W., Yu, J., and Lanza, G. (2003). *Genetic Programming IV: Routine Human-Competitive Machine Intelligence.* Kluwer Academic Publishers.

Kwiat, P., Weinfurter, H., Herzog, T., Zeilinger, A., and Kasevich, M. (1995). Interaction-free quantum measurements. *Physical Review Letters*, 74:4763–4766.

Landauer, R. (1999). Information is inevitably physical. In Hey, J. G., editor, *Feynman and Computation: Exploring the Limits of Computers*, pages 77–92. Perseus Books, Reading, MA.

Langdon, W. B. (1998). *Genetic Programming and Data Structures: Genetic Programming + Data Structures = Automatic Programming!* Kluwer, Boston.

Langdon, W. B., Soule, T., Poli, R., and Foster, J. A. (1999). The evolution of size and shape. In Spector, L., Langdon, W. B., O'Reilly, U.-M., and Angeline, P. J., editors, *Advances in Genetic Programming 3*, chapter 8, pages 163–190. MIT Press, Cambridge, MA, USA.

Leier, A. and Banzhaf, W. (2003a). Evolving Hogg's quantum algorithm using linear-tree GP. In Cantú-Paz, E., Foster, J. A., Deb, K., Davis, D., Roy, R., O'Reilly, U.-M., Beyer, H.-G., Standish, R., Kendall, G., Wilson, S., Harman, M., Wegener, J., Dasgupta, D., Potter, M. A., Schultz, A. C., Dowsland, K., Jonoska, N., and Miller, J., editors, *Genetic and Evolutionary Computation – GECCO-2003*, volume 2723 of *LNCS*, pages 390–400, Chicago. Springer-Verlag.

Leier, A. and Banzhaf, W. (2003b). Exploring the search space of quantum programs. In Sarker, R. et al., editors, *Proc. 2003 Congress on Evolutionary Computation (CEC'03), Canberra*, volume 1, pages 170–177, Piscataway NJ. IEEE Press.

Luke, S. (2000). *Issues in Scaling Genetic Programming: Breeding Strategies, Tree Generation, and Code Bloat.* PhD thesis, Department of Computer Science, University of Maryland, A. V. Williams Building, University of Maryland, College Park, MD 20742 USA.
http://www.cs.gmu.edu/~sean/papers/thesis2p.pdf.

Luke, S. and Spector, L. (1998). A revised comparison of crossover and mutation in genetic programming. In Koza, J. R., Banzhaf, W., Chellapilla, K., Deb, K., Dorigo, M., Fogel, D. B., Garzon, M. H., Goldberg, D. E., Iba, H., and Riolo, R., editors, *Genetic Programming 1998: Proceedings of the Third Annual Conference*, pages 208–213, University of Wisconsin, Madison, Wisconsin, USA. Morgan Kaufmann.

Massey, P., Clark, J., and Stepney, S. (2004). Evolving quantum circuits and programs through genetic programming. In Deb, K., Poli, R., Spector, L., Thierens, D., Beyer, H.-G., Tettamanzi, A., Lanzi, P. L., Tyrrell, A., Foster, J., Banzhaf, W., Holland, O., Floreano, D., Burke, E., Harman, M., Darwen, P., and Dasgupta, D., editors, *Genetic and Evolutionary Computation – GECCO-2004.* Springer-Verlag.

McCarthy, J., Levin, M., et al. (1966). *LISP 1.5 Programmer's Manual.* MIT.

Milburn, G. J. (1997). *Schrödinger's Machines: The Quantum Technology Reshaping Everyday Life.* W. H. Freeman and Company, New York.

Mitchell, M. (1996). *An Introduction to Genetic Algorithms.* MIT Press.

Montana, D. J. (1993). Strongly typed genetic programming. BBN Technical Report #7866, Bolt Beranek and Newman, Inc., 10 Moulton Street, Cambridge, MA 02138, USA.

Nielsen, M. A. and Chuang, I. L. (2000). *Quantum Computation and Quantum Information.* Cambridge University Press, Cambridge.

Obenland, K. and Despain, A. (1998). A parallel quantum computer simulator. `http://arxiv.org/quant-ph/9804039`.

O'Neill, M. and Ryan, C. (2003). *Grammatical Evolution: Evolutionary Automatic Programming in a Arbitrary Language*, volume 4 of *Genetic programming.* Kluwer Academic Publishers.

Penrose, R. (1989). *The Emperor's New Mind: concerning computers, minds, and the laws of physics.* Oxford University Press.

Penrose, R. (1997). *The Large, the Small and the Human Mind.* Cambridge University Press.

Perkis, T. (1994). Stack-based genetic programming. In *Proceedings of the 1994 IEEE World Congress on Computational Intelligence*, volume 1, pages 148–153, Orlando, Florida, USA. IEEE Press.

Perkowski, M., Lukac, M., Pivtoraiko, M., Kerntopf, P., Folgheraiter, M., Lee, D., Kim, H., Kim, H., Hwangboo, W., Kim, J.-W., and Choi, Y. (2003). A hierarchical approach to computer aided design of quantum circuits. In *Proceedings of 6th International Symposium on Representations and Methodology of Future Computing Technology, RM 2003*, pages 201–209. `http://www.ee.pdx.edu/~mperkows/=PUBLICATIONS/PDF-2003/Perkowski.pdf`.

Polito, J., Daida, J., and Bersano-Begey, T. F. (1997). Musica ex machina: Composing 16th-century counterpoint with genetic programming and symbiosis. In Angeline, P. J., Reynolds, R. G., McDonnell, J. R., and Eberhart, R., editors, *Evolutionary Programming VI: Proceedings of the Sixth Annual Conference on Evolutionary Programming*, volume 1213 of *Lecture Notes in Computer Science*, Indianapolis, Indiana, USA. Springer-Verlag.

Racine, A., Schoenauer, M., and Dague, P. (1998). A dynamic lattice to evolve hierarchically shared subroutines: DL'GP. In Banzhaf, W., Poli, R., Schoenauer, M., and Fogarty, T. C., editors, *Proceedings of the First European Workshop on Genetic Programming*, volume 1391 of *LNCS*, pages 220–232, Paris. Springer-Verlag.

Rieffel, E. and Polak, W. (2000). An introduction to quantum computing for non-physicists. `http://arxiv.org/quant-ph/9809016`.

Riolo, R. L. and Worzel, B. (2003). *Genetic Programming Theory and Practice.* Kluwer, Boston, MA, USA.

Roberts, S. C., Howard, D., and Koza, J. R. (2001). Evolving modules in genetic programming by subtree encapsulation. In Miller, J. F., Tomassini, M., Lanzi, P. L., Ryan, C., Tettamanzi, A. G. B., and Langdon, W. B., editors, *Genetic Programming, Proceedings of EuroGP'2001*, volume 2038 of *LNCS*, pages 160–175, Lake Como, Italy. Springer-Verlag.

Robinson, A. (2001). Genetic programming: Theory, implementation, and the evolution of unconstrained solutions. Division III thesis, Hampshire College. `http://hampshire.edu/lspector/robinson-div3.pdf`.

Rylander, B., Soule, T., Foster, J., and Alves-Foss, J. (2001). Quantum evolutionary programming. In Spector, L., Goodman, E. D., Wu, A., Langdon, W. B., Voigt, H.-M., Gen, M., Sen, S., Dorigo, M., Pezeshk, S., Garzon, M. H., and Burke,

E., editors, *Proceedings of the Genetic and Evolutionary Computation Conference (GECCO-2001)*, pages 1005–1011, San Francisco, California, USA. Morgan Kaufmann.

Schmidhuber, J. (1987). Evolutionary principles in self-referential learning. on learning now to learn: The meta-meta-meta...-hook. Diploma thesis, Technische Universitat Munchen, Germany.

Schumacher, C., Vose, M. D., and Whitley, L. D. (2001). The no free lunch and problem description length. In Spector, L., Goodman, E. D., Wu, A., Langdon, W. B., Voigt, H.-M., Gen, M., Sen, S., Dorigo, M., Pezeshk, S., Garzon, M. H., and Burke, E., editors, *Proceedings of the Genetic and Evolutionary Computation Conference (GECCO-2001)*, pages 565–570, San Francisco, California, USA. Morgan Kaufmann.

Shapiro, J. H. and Hirota, O., editors (2003). *Proceedings of the 6th International Conference on Quantum Communication, Measurement, and Computing.* Rinton Press, Princeton, New Jersey.

Shor, P. W. (1994). Algorithms for quantum computation: Discrete logarithms and factoring. In Goldwasser, S., editor, *Proceedings of the 35th Annual Symposium on Foundations of Computer Science.* IEEE Computer Society Press.

Shor, P. W. (1998). Quantum computing. *Documenta Mathematica*, Extra Volume ICM:467–486. `http://east.camel.math.ca/EMIS/journals/DMJDMV/xvol-icm / 00/Shor.MAN.ps.gz`.

Spector, L. (1996). Simultaneous evolution of programs and their control structures. In Angeline, P. J. and Kinnear, Jr., K. E., editors, *Advances in Genetic Programming 2*, chapter 7, pages 137–154. MIT Press, Cambridge, MA, USA.

Spector, L. (2001). Autoconstructive evolution: Push, pushGP, and pushpop. In Spector, L., Goodman, E. D., Wu, A., Langdon, W. B., Voigt, H.-M., Gen, M., Sen, S., Dorigo, M., Pezeshk, S., Garzon, M. H., and Burke, E., editors, *Proceedings of the Genetic and Evolutionary Computation Conference (GECCO-2001)*, pages 137–146, San Francisco, California, USA. Morgan Kaufmann.

Spector, L. (2002). Adaptive populations of endogenously diversifying pushpop organisms are reliably diverse. In Standish, R. K., Bedau, M. A., and Abbass, H. A., editors, *Proceedings of Artificial Life VIII, the 8th International Conference on the Simulation and Synthesis of Living Systems*, pages 142–145. The MIT Press.

Spector, L. (2003). An essay concerning human understanding of genetic programming. In Riolo, R. L. and Worzel, B., editors, *Genetic Programming Theory and Practice*, chapter 2, pages 11–24. Kluwer.

Spector, L. and Alpern, A. (1994). Criticism, culture, and the automatic generation of artworks. In *Proceedings of Twelfth National Conference on Artificial Intelligence*, pages 3–8, Seattle, Washington, USA. AAAI Press/MIT Press.

Spector, L., Barnum, H., and Bernstein, H. J. (1998). Genetic programming for quantum computers. In Koza, J. R., Banzhaf, W., Chellapilla, K., Deb, K., Dorigo, M., Fogel, D. B., Garzon, M. H., Goldberg, D. E., Iba, H., and Riolo, R., editors, *Genetic Programming 1998: Proceedings of the Third Annual Conference*, pages 365–373, University of Wisconsin, Madison, Wisconsin, USA. Morgan Kaufmann.

Spector, L., Barnum, H., Bernstein, H. J., and Swami, N. (1999a). Finding a better-than-classical quantum AND/OR algorithm using genetic programming. In Angeline, P. J., Michalewicz, Z., Schoenauer, M., Yao, X., and Zalzala, A., editors, *Proceedings of the Congress on Evolutionary Computation*, volume 3, pages 2239–2246, Mayflower Hotel, Washington D.C., USA. IEEE Press.

Spector, L., Barnum, H., Bernstein, H. J., and Swamy, N. (1999b). Quantum computing applications of genetic programming. In Spector, L., Langdon, W. B., O'Reilly, U.-M., and Angeline, P. J., editors, *Advances in Genetic Programming 3*, chapter 7, pages 135–160. MIT Press, Cambridge, MA, USA.

Spector, L. and Bernstein, H. J. (2003). Communication capacities of some quantum gates, discovered in part through genetic programming. In Shapiro, J. H. and Hirota, O., editors, *Proceedings of the Sixth International Conference on Quantum Communication, Measurement, and Computing (QCMC)*, pages 500–503. Rinton Press.

Spector, L. and Klein, J. (2002). Evolutionary dynamics discovered via visualization in the BREVE simulation environment. In Smith, T., Bullock, S., and Bird, J., editors, *Beyond Fitness: Visualising Evolution — Work. of Artificial Life VIII: 8th Int. Conf. Simulation and Synthesis of Living Systems.*

Spector, L., Klein, J., Perry, C., and Feinstein, M. (2003a). Emergence of collective behavior in evolving populations of flying agents. In Cantú-Paz, E., Foster, J. A., Deb, K., Davis, D., Roy, R., O'Reilly, U.-M., Beyer, H.-G., Standish, R., Kendall, G., Wilson, S., Harman, M., Wegener, J., Dasgupta, D., Potter, M. A., Schultz, A. C., Dowsland, K., Jonoska, N., and Miller, J., editors, *Genetic and Evolutionary Computation - GECCO-2003*, volume 2723 of *LNCS*, pages 61–73, Chicago. Springer-Verlag.

Spector, L., Langdon, W. B., O'Reilly, U.-M., and Angeline, P. J., editors (1999c). *Advances in Genetic Programming 3.* MIT Press, Cambridge, MA, USA.

Spector, L., Perry, C., and Klein, J. (2003b). Push 2.0 programming language description. Technical report, School of Cognitive Science, Hampshire College. `http://hampshire.edu/lspector/push2-description.html`.

Spector, L. and Robinson, A. (2002a). Genetic programming and autoconstructive evolution with the push programming language. *Genetic Programming and Evolvable Machines*, 3(1):7–40.

Spector, L. and Robinson, A. (2002b). Multi-type, self-adaptive genetic programming as an agent creation tool. In Barry, A. M., editor, *GECCO 2002: Proceedings of the Bird of a Feather Workshops, Genetic and Evolutionary Computation Conference*, pages 73–80, New York. AAAI.
`http://hampshire.edu/lspector/pubs/ecomas2002-spector-toappear.pdf`.

Spector, L. and Stoffel, K. (1996a). Automatic generation of adaptive programs. In Maes, P., Mataric, M. J., Meyer, J.-A., Pollack, J., and Wilson, S. W., editors, *Proceedings of the Fourth International Conference on Simulation of Adaptive Behavior: From animals to animats 4*, pages 476–483, Cape Code, USA. MIT Press.

Spector, L. and Stoffel, K. (1996b). Ontogenetic programming. In Koza, J. R., Goldberg, D. E., Fogel, D. B., and Riolo, R. L., editors, *Genetic Programming 1996: Proceedings of the First Annual Conference*, pages 394–399, Stanford University, CA, USA. MIT Press.

Steane, A. (1998). Quantum computing. *Reports on Progress in Physics*, 61:117–173. `http://xxx.lanl.gov/abs/quant-ph/9708022`.

Steele Jr., G. L. (1984). *Common LISP: The Language.* Digital Press, Burlington, Mass.

Stoffel, K. and Spector, L. (1996). High-performance, parallel, stack-based genetic programming. In Koza, J. R., Goldberg, D. E., Fogel, D. B., and Riolo, R. L., editors, *Genetic Programming 1996: Proceedings of the First Annual Conference*, pages 224–229, Stanford University, CA, USA. MIT Press.

Surkan, A. J. and Khuskivadze, A. (2001). Evolution of quantum algorithms for computer of reversible operators. In Stoica, A., Lohn, J., Katz, R., Keymeulen, D., and Zebulum, R. S., editors, *The 2002 NASA/DoD Conference on Evolvable Hardware*, pages 186–187, Long Beach, California. Jet Propulsion Laboratory, California Institute of Technology, IEEE Computer Society.

Tchernev, E. (1998). Forth crossover is not a macromutation? In Koza, J. R., Banzhaf, W., Chellapilla, K., Deb, K., Dorigo, M., Fogel, D. B., Garzon, M. H., Goldberg, D. E., Iba, H., and Riolo, R., editors, *Genetic Programming 1998: Proceedings of the Third Annual Conference*, pages 381–386, University of Wisconsin, Madison, Wisconsin, USA. Morgan Kaufmann.

Udrescu-Milosav, M. (2003). *Quantum Algorithms Implementation: Circuit Design Principles and Entanglement Analysis.* PhD thesis, University Politehnica Timisoara, Romania.

Vaidman, L. (1996). Interaction-free measurements.
`http://arxiv.org/quant-ph/9610033`.

van Dam, W., Hallgren, S., and Ip, L. (2002). Quantum algorithms for some hidden shift problems.
`http://arxiv.org/quant-ph/0211140`.

van Dam, W. and Seroussi, G. (2002). Efficient quantum algorithms for estimating Gauss sums.
`http://arxiv.org/quant-ph/0207131`.

Viamontes, G. F., Markov, I. L., and Hayes, J. P. (2003). Improving gate-level simulation of quantum circuits.
`http://arxiv.org/quant-ph/0309060`.

Viamontes, G. F., Rajagopalan, M., Markov, I. L., and Hayes, J. P. (2002). Gate-level simulation of quantum circuits.
`http://arxiv.org/quant-ph/0208003`.

Whitley, D. (1999). A free lunch proof for gray versus binary encodings. In Banzhaf, W., Daida, J., Eiben, A. E., Garzon, M. H., Honavar, V., Jakiela, M., and Smith, R. E., editors, *Proceedings of the Genetic and Evolutionary Computation Conference*, volume 1, pages 726–733, Orlando, Florida, USA. Morgan Kaufmann.

Wiles, J. and Tonkes, B. (2002). Visualisation of hierarchical cost surfaces for evolutionary computation. In *Proceedings of the 2002 Congress on Evolutionary Computation*, pages 157–162.

Williams, C. P. and Clearwater, S. H. (1998). *Explorations in Quantum Computing.* Springer-Verlag, New York.

Williams, C. P. and Gray, A. G. (1999). Automated design of quantum circuits. In Williams, C. P., editor, *Quantum Computing and Quantum Communications: First NASA International Conference, QCQC'98*, number 1509 in Lecture Notes in Computer Science, LNCS, pages 113–125, Palm Springs, California, USA. Springer-Verlag.

Wolpert, D. H. and Macready, W. G. (1997). No free lunch theorems for optimization. *IEEE Trans. on Evolutionary Computation*, 1(1):67–82.

Woodward, J. R. and Neil, J. R. (2003). No free lunch, program induction and combinatorial problems. In Ryan, C., Soule, T., Keijzer, M., Tsang, E., Poli, R., and Costa, E., editors, *Genetic Programming, Proceedings of EuroGP'2003*, volume 2610 of *LNCS*, pages 479–488, Essex. Springer-Verlag.

About the Author

Lee Spector is Dean of the School of Cognitive Science and Professor of Computer Science at Hampshire College. He received a B.A. in Philosophy from Oberlin College in 1984, and a Ph.D. from the Department of Computer Science at the University of Maryland in 1992.

Dr. Spector teaches and conducts research in computer science, artificial intelligence, and artificial life. He recently received the highest honor bestowed by the National Science Foundation for excellence in both teaching and research, the *NSF Director's Award for Distinguished Teaching Scholars.* His areas of interest include genetic and evolutionary computation, quantum computation, planning in dynamic environments, artificial intelligence education, artificial intelligence and neuropsychology, and artificial intelligence in the arts.

Dr. Spector has produced over 50 professional scientific publications and serves as an Associate Editor for the journal *Genetic Programming and Evolvable Machines*, published by Kluwer. He was Editor-in-Chief for the Proceedings of the 2001 GECCO (Genetic and Evolutionary Computation) conference and he has been the Program Chair for the Genetic Programming and Evolvable Hardware tracks of other GECCO conferences. He was lead editor for *Advances in Genetic Programming*, Volume 3, published by MIT Press.

About the Author

Index